ACID POLITICS

ACID POLITICS: ENVIRONMENTAL AND ENERGY POLICIES IN BRITAIN AND GERMANY

Sonja Boehmer-Christiansen
and
Jim Skea

Belhaven Press
(a division of Pinter Publishers)
London and New York

© Sonja Boehmer-Christiansen and Jim Skea, 1991

First published in Great Britain in 1991 by
Belhaven Press (a division of Pinter Publishers),
25 Floral Street, London WC2E 9DS

British Library Cataloguing in Publication Data
A CIP catalogue record for this book is available from the
British Library

ISBN 1 85293 116 7

For enquiries in North America please contact PO Box 197,
Irvington, NY 10533

Library of Congress Cataloging in Publication Data

Boehmer-Christiansen, Sonja.
    Acid Politics: environmental and energy policies in Britain and
  West Germany/ by Sonja Boehmer-Christiansen and Jim Skea.
      p.   cm.
    Includes index.
    ISBN 1-85293-116-7
    1. Environmental policy—Great Britain. 2. Energy policy—Great
  Britain. 3. Environmental policy—Germany (West) 4.Energy policy
  —Germany (West) I. Skea, Jim. II. Title.
HC260.E5B64   1991
363.7'056'0941—dc20   90–48870
                                                                        CIP

Typeset by Florencetype Ltd, Kewstoke, Avon
Printed and bound by Biddles Ltd.

Contents

List of figures

List of tables

Acknowledgements

This book is based on research which was supported primarily by the Joint Research Committee (JRC) of the Economic and Social Research Council (ESRC) and the Science and Engineering Research Council (SERC) under a grant for work on 'Acid Deposition and the European Energy Sector' during the period 1985–87. It was possible to maintain the work beyond 1987 through a further grant from the JRC on 'Environmental Regulation, Technical Change and the Energy Sector' and through associated support from the Anglo-German Foundation and the ESRC's Designated Research Centre in Science, Technology and Energy Policy at the Science Policy Research Unit (SPRU), University of Sussex.

The work has benefited enormously from interviews with present and former employees of a variety of organizations to whom we owe a large debt of gratitude. In the UK, the organizations included the former Central Electricity Generating Board, the Confederation of British Industry, the Department of the Environment, Earth Resources Research, HM Inspectorate of Pollution, ICI and the Institution of Professional Civil Servants.

The corresponding organizations in Germany included the Association of German Electricity Producers, the Association of Large Power Plant Operators, the Federal Environment Ministry, the Federal Environment Office (UBA), the Federation of German Industry and the electricity utility RWE. Meetings were also held with officials of the Land Governments of Bavaria and North Rhine-Westphalia.

At the international level, we are grateful for discussions with the Environment and Energy Directorates-General of the European Commission, with the UK Permanent Representation in Brussels and with the West German and British Embassies in London and Bonn respectively.

A number of those interviewed also commented on drafts of sections of the book. We are especially grateful to them for removing some early misconceptions and pointing us in the right direction.

The support of many academic colleagues was invaluable. SPRU colleagues, John Surrey in particular, read and commented on parts of the book. Tim O'Riordan of the School of Environmental Sciences at the University of East Anglia, Roger Williams of the Department of Government at the University of Manchester, Mike Chadwick of the Beijer Institute Centre at the University of York, Helmut Weidner of the Wissenschaftszentrum Berlin and colleagues at the Institute for International Public Law at the University of Munich made suggestions and provided fresh insights.

The monthly Environmental Data Services – ENDS report – features strongly in the chapter notes and deserves a particular mention. ENDS is an indispensible resource for anyone carrying out environmental research in the UK and this has proved to be as true of the acid-rain issue as of any other.

In co-writing a book which covers an issue as contentious as acid rain, we have necessarily debated and attempted to reconcile divergent views which have been strongly held, both by ourselves and by those to whom we talked. At the end of the day, the opinions expressed, and any errors of fact or interpretation, remain, to invoke British government practice, the collective responsibility of the two authors.

List of English Abbreviations

ACAI	Alkali and Clean Air Inspectorate (later the IAPI)
AFBC	atmospheric fluidized bed combustion
AGR	advanced gas-cooled reactor
BATNEEC	best available techniques not entailing excessive cost
BCC	British Coal Corporation (formerly the National Coal Board)
bn	billion
BPEO	best practicable environmental option
BPM	best practicable means
BWR	boiling water reactor
CCGT	combined cycle gas turbine
CDEP	Central Directorate on Environmental Pollution (part of the DoE)
CEC	Commission of the European Communities
CEGB	Central Electricity Generating Board
CENE	Commission on Energy and the Environment
CFC	chlorofluorocarbon
CND	Campaign for Nuclear Disarmament
CO_2	carbon dioxide
CPRE	Council for the Protection of Rural England
DANW	Directorate of Air, Noise and Waste (part of the DoE)
DEA	Department of Economic Affairs
DEn	Department of Energy
DM	deutschmark
DoE	Department of the Environment
DTI	Department of Trade and Industry
DTp	Department of Transport
EB	Executive Board (of the LRTAP Convention)
EC	European Community
EMEP	European Monitoring and Evaluation Programme
ESI	electricity supply industry
FBC	fluidized bed combustion
FCO	Foreign and Commonwealth Office
FGD	flue gas desulphurization
FoE	Friends of the Earth
FRG	Federal Republic of Germany
GDP	gross domestic product
HMIP	Her Majesty's Inspectorate of Pollution
HMIPI	Her Majesty's Industrial Pollution Inspectorate (Scotland)
HSC	Health and Safety Commission
HSE	Health and Safety Executive
IAPI	Industrial Air Pollution Inspectorate (later part of HMIP)
IGCC	integrated gasification combined cycle
IIASA	International Institute for Applied Systems Analysis
IPC	integrated pollution control
IPCC	Intergovernmental Panel on Climate Change
IUCN	International Union for the Conservation of Nature

kW	kilowatt
LCP	large combustion plant
LRTAP	long-range transboundary air pollution
m	million
MAFF	Ministry of Agriculture, Forestry and Fisheries
mg	milligrammes
MINIS	Management Information System for Ministers
MP	Member of Parliament
mtce	million tonnes of coal equivalent
MWe	megawatt (thousand kW) electrical
MWth	megawatt (thousand kW) thermal
NCB	National Coal Board (later BCC)
NCC	Nature Conservancy Council
NIE	Northern Ireland Electricity
NII	Nuclear Installations Inspectorate
Nm^3	normal cubic metres
NO_x	nitrogen oxides
NRA	National Rivers Authority
NSCA	National Society for Clean Air
NSHEB	North of Scotland Hydro-Electric Board
NUM	National Union of Mineworkers
OECD	Organisation for Economic Cooperation and Development
PFBC	pressurized fluidized bed combustion
ppm	parts per million
PPP	polluter pays principle
PSBR	public sector borrowing requirement
PWR	pressurized water reactor
R&D	research and development
RAINS	reversing acidification in Norwegian soils
RCEP	Royal Commission on Environmental Pollution
REC	Regional Electricity Company
RSPB	Royal Society for the Protection of Birds
SCR	selective catalytic reduction
SDP	Social Democratic Party
SEG	System Environment Group (formerly part of CEGB)
SO_2	sulphur dioxide
SSEB	South of Scotland Electricity Board
TCPA	Town and Country Planning Association
TFP	transfrontier pollution
TPRD	Technology Planning and Research Division (part of CEGB)
TWh	terawatt hours (billion kWh)
UDM	Union of Democratic Mineworkers
UNECE	UN Economic Commission for Europe
UNEP	UN Environment Programme
WHO	World Health Organisation
WMO	World Meteorological Organisation
WSL	Warren Spring Laboratory
WWF	Worldwide Fund for Nature (formerly the World Wildlife Fund)

List of German Abbreviations

AA	Auswärtiges Amt	Foreign Ministry
AG	Aktiengesellschaft	limited liability company
BBU	Bund der Bürgerinitiativen für Umweltschutz	Federation of Citizens Action Groups for the Environment
BDI	Bund der Deutschen Industrie	Federation of German Industry
BImSchG	Bundesimmissionsschutzgesetz	Federal Air Quality Protection Law
BImSchV	Bundesimmissionsschutz-verordnung	regulation issued under the BImSchG
BMELF	Bundesministerium für Ernährung, Landwirtschaft und Forsten	Federal Ministry for Nutrition, Agriculture and Forestry
BMF	Bundesministerium für Finanzen	Federal Ministry of Finance
BMFT	Bundesministerium für Forschung und Technologie	Federal Ministry for Research and Technology
BMG	Bundesministerium für Gesundheitswesen	Federal Health Ministry
BMI	Bundesministerium des Innern	Federal Interior Ministry
BMU	see BMUNRS	
BMUNRS	Bundesministerium für Umwelt, Naturschutz und Reaktorsicherheit	Federal Ministry for Environment, Nature Protection and Reactor Safety
BMWi	Bundesministerium für Wirtschaft	Federal Ministry for Economics
CDU	Christlich Demokratische Union	Christian Democratic Union
CSU	Christlich Soziale Union	Christian Social Union
DIW	Deutsches Institut für Wirtschaftsforschung	German Institute for Economic Research
DVG	Deutsche Verbund Gesellschaft	association of electric utilities controlling transmission network
EnWiG	Energiewirtschaftgesetz	Energy Sector Law
FDP	Freie Demokratische Partei	Free Democratic Party
GAZ	Grüne Aktion Zukunft	Green Action for the Future
GeWO	Gewerbeordnung	General Trade Regulation
GFAVo	Grossfeurungsanlagenverordnung	large combustion plant regulation
IPA	Interparliamentarische Arbeitsgemeinschaft für natur-gemässe Wirtschaftsweise	Interparliamentary Working Group for Nature-Appropriate Economic Activities
KWU	Kraftwerk Union	nuclear power plant construction company
LRG	Luftreinhaltegesetz	Federal Air Purity Law
NRW	Nordrhein-Westfalen	Land of North Rhine-Westphalia
NWK	Nordwestdeutsche Kraftwerke	electric utility
RBK	Rheinische Braunkohlenwerke	lignite mining subsidiary of RWE

RSU	Rat der Sachverständigen für Umweltfragen	Advisory Council on Environmental Questions
RWE	Rheinisch Westfälische Elektrizitätswerke	NRW-based electric utility
SPD	Sozialdemokratische Partei Deutschland	Social Democratic Party
STEAG	Steinkohlen Elektrizitäts AG	Hard Coal Electricity Company
TA(Luft)	Technische Anleitung zur Reinhaltung der Luft	technical guidelines for maintenance of pure air
TUV	Technischer Überwachungsverein	Technical Inspection Office
UBA	Umweltbundesamt	Federal Environment Office
VDEW	Vereinigung der Deutschen Elektrizitätswerke	Association of German Electricity Producers
VDI	Verein Deutscher Ingenieure	Association of German Engineers
VKU	Verband Kommunaler Unternehmen	Association of Municipal Undertakings

German words used in book

Bundesrat	Federal Council (upper house of Parliament)
Bundestag	lower house of Parliament
Bürgerinitiativen	citizens action groups
Entsorgung	provision for waste disposal
Fundi	fundamentalist – uncompromising Green
Immission	concentration of pollutants in atmosphere
Jahrhundertvertrag	century contract for hard coal
Kohlepfennig	'coal penny' – hard coal subsidy
Land, Länder	state(s) of the Federal Republic
Politik	politics or policy
Realo	realist – pragmatic Green
Schadstoff	pollutant
Stand der Technik	state-of-the-art technology
Umwelt	environment
Umweltschutz	environmental protection
Verordnung	regulation
Verschmutzung	pollution
Verwaltungsvorschrift	administrative directive
Vollzugsdefizit	implementation gap
Vorsorge	precaution, anticipation
Wald	forest
Waldschäden	forest damages
Waldsterben	forest dieback

PART I
The wider setting

1 Introduction

This book is a case study of the efforts made during the 1980s to resolve a major international environmental problem, acid rain, in two West European countries, the UK and the Federal Republic of Germany.

The acid rain problem itself has been the subject of many publications.[1] This book differs from many of those previously published in that it is concerned almost exclusively with the social, political, administrative and economic pressures which led to policy-making taking the particular course that it did. It is not the primary concern to describe the science of acid rain or the technical means for reducing it. The policy-making processes themselves in the UK and the Federal Republic of Germany (FRG) are the focus of attention.

At the start of the 1990s, the major problem facing environmental policy makers is global climate change. Serious consideration is now being given to a Climate Convention which would require the co-ordination of policy responses at a global level. Why then, when another major environmental issue is looming, is it worth while re-examining the acid rain problem in any depth?

The answer lies partly in the fact that the acid-rain problem has yet to be completely resolved in Europe and North America. Only now is a more complete scientific understanding beginning to emerge about the extent of acid rain effects and the magnitude of the remedial measures which would be required if further environmental damage were to be prevented, never mind existing damage reversed. Equally, although a broad measure of international agreement on the type of steps required to combat acid rain has been reached, policy measures are only now being put into effect in many countries, the UK included.

In addition, there are many features of the policy-making processes experienced during the resolution of the acid-rain problem which may bear lessons for future efforts made in response to climate change or other global and regional environmental issues:

1 energy production and transformation, undertaken by large, politically influential companies, often of a monopolistic or quasi-monopolistic nature, is one of the main economic sectors potentially affected by both acid rain and climate change policies;
2 acid rain, while not a truly global problem, is an issue at a continental level and has hence had to be resolved through international diplomacy and co-operation;

3 responses have generally required major investments in abatement equipment which tend to be strongly resisted by interested parties until government
   plays an active regulatory role;
4 policy responses may significantly alter incentives for particular technology
   and fuel choices; and
5 particular countries and companies may perceived economic advantage in
   ambitious policy responses which others may resist as major economic
   threats.

Britain and Germany

The fundamental reason for selecting Britain and Germany as the objects of a
comparative case study is the widely divergent policy stances which were taken
by the two countries on the acid rain issue. This divergence is reflected in the
understanding and assessment of acid rain damage, the anticipated impacts of
abatement options and even in the objectives of environmental policy itself.

A further reason for selecting the UK and the FR Germany is that the
European Community (EC) became one of the main fora within which policy
tensions became resolved. In many ways, the UK and the FR Germany
represent the poles of the bitter debate which took place between the
Commission of the European Communities (CEC) and the Member States
during the period 1983 and 1988.

While neither country has been the victim of large-scale 'classic' acid-rain
damage, lifeless lakes and streams, each has suffered in its own way. Damage to
German forests linked to acid rain and other types of air-pollution effects was a
major policy stimulus in the early 1980s. Parts of Britain have suffered more
limited damage to lakes and streams while British trees have been as badly
affected as German ones, although less public anxiety has been occasioned.

The UK and FR Germany are major emitters of sulphur dioxide (SO_2) and
nitrogen oxides (NO_x), the two main man-made gases giving rise to acid rain.
Both are major industrial countries with large-scale domestic energy industries
and electricity sectors heavily reliant on the use of indigenous coal.

There are considerable geographical differences between Britain and
Germany and this has significantly affected not only the perception of acid rain
but also the objective experience of acid deposition. While Germany, at the
heart of continental Europe, shares its atmosphere with many neighbours, some
relatively clean, others to the east decidedly not, Britain is a wind-swept
off-shore island little affected by atmospheric pollution originating abroad.
However, the UK and the FR Germany are themselves the largest polluters in
Western Europe and winds spread the SO_2 and NO_x originating from British
and German chimneys over a wide continental area.

Tables 1.1 and 1.2 show how British and German emissions of SO_2 are
estimated to have contributed to deposition at home and in Scandinavia in the
years 1980 and 1988 respectively. These tables bring out, in numerical form,
some of the underlying factors which have helped to shape British and German
(and Scandinavian) attitudes to acid rain:

Table 1.1   Sulphur budgets for Britain and Germany in 1980
(ktonnes sulphur dioxide per year)

Emission receivers	Emission sources							
	UK	FRG	Sweden	Norway	Other Europe	Total Europe	Unknown sources	Total
UK	1620	50	–	–	190	1860	173	2033
FRG	173	1346	–	–	1920	3439	192	3631
Sweden	84	84	199	48	446	862	271	1133
Norway	96	48	24	19	247	434	178	612
Other Europe	1020	1939	134	17	34250	37631	5148	42509
Total Europe	2993	3468	358	84	37053	43956	5962	49918
Residuals	2126	163	192	67	11585	14134		
Total emissions	5119	3631	550	151	48638	58090		

Notes: (1) 'Other Europe' includes the Soviet Union; (2) 'Residuals' includes deposition at sea, outside Europe and contributions to 'unknown sources'.
Sources: EMEP.

Table 1.2   Sulphur budgets for Britain and Germany in 1988
(ktonnes sulphur dioxide per year)

Emission receivers	Emission sources							
	UK	FRG	Sweden	Norway	Other Europe	Total Europe	Unknown sources	Total
UK	1080	10	–	–	460	1550	144	1694
FRG	86	552	–	–	2228	2866	160	3026
Sweden	40	28	68	6	576	718	226	944
Norway	50	18	6	18	270	362	148	510
Other Europe	378	532	24	–	14146	15080	2826	17906
Total Europe	1634	1140	98	24	17680	20576	3504	24080
Residuals	2146	750	122	48	18214	21280		
Total emissions	3780	1890	220	72	35894	41856		

Notes: (1) 'Other Europe' includes the European sector of the Soviet Union; (2) 'Residuals' includes deposition at sea, outside Europe and contributions to 'unknown sources'; (3) differences between Tables 1.1 and 1.2 reflect modelling improvements as well as changes in actual deposition patterns.
Sources: EMEP.

1   the UK is responsible for most of the SO$_2$ deposition within its own borders: the FR Germany is not;
2   as an off-shore island, the destination of much of the UK's emissions is not certain: a large proportion is believed to be deposited harmlessly at sea. However, most German emissions are deposited in continental Europe;

3 both the UK and the FR Germany are, or have been, bigger contributors
   to Norwegian deposition than has Norway itself: while contributions to
   deposition in Sweden have been less than those of Sweden itself, they are of
   the same order of magnitude;
4 between 1980 and 1988, the FR Germany, Sweden and Norway have
   reduced their SO_2 emissions by 48 per cent, 60 per cent and 52 per cent
   respectively. However, the British reduction has been only 26 per cent.

Emissions of NO_x from stationary sources have also declined faster in the FR
Germany than in the UK.

These figures give, therefore, in a nut-shell, the reasons for Scandinavian
resentment, German receptiveness to the idea of emission controls and British
reluctance to follow suit.

The story of acid rain in Britain and Germany

The story of acid-rain policy-making in Britain and Germany is described at
greater length in Chapters 10 and 11. However, it is worth while describing at
this point, in outline, the pattern of events which involved the two countries
during the 1980s.

The FR Germany had made considerable progress in developing a modern,
Federally based environmental control system during the 1970s. In 1974, an
air-quality law was enacted and a Federal body, the Umweltbundesamt (UBA
– the Federal Environment Office), was set up to advise the Government in
Bonn on environmental matters. Although interest waned somewhat during the
latter part of the decade, UBA helped to draft legislation in 1978 which was to
form the basis for the wider European effort to clean up acid rain in the 1980s.

Until 1982, both the UK and the FR Germany had resisted international
pressure from the Scandinavians to make significant reductions in acid
emissions. Both countries had fought against specific SO_2 emission reduction
requirements being built into the 1979 Geneva Convention on Long Range
Transboundary Air Pollution (LRTAP). However, in 1982, Germans became
aware of widespread damage to their forests, the cause of which was believed to
be acid rain. Given the importance of forests in both the German economy and
culture, there was widespread approval of steps taken to cut drastically
emissions of SO_2 and NO_x from power stations and other sources. A
considerable amount of debate took place about the exact form which controls
should take given the important implications for the energy industries which are
largely in private ownership.

Most of the coal-fired power stations in the FR Germany were fitted with
flue gas desulphurization (FGD) equipment over a remarkably short period of
time, between 1983 and 1988. A major programme of retrofitting selective
catalytic reduction (SCR) technology to all hard coal-fired power stations, with
the objective of cutting NO_x emissions, has taken a little longer.

The enactment of the Grossfeuerungsanlagenverordnung (GFAVo – Large
Combustion Plant Regulation) enabled the FR Germany to reverse its international position on acid rain and to press for tighter international controls. It

did so most vigorously, and, from the point of view of the UK, most persuasively, through the machinery of the EC.

In December 1983, the CEC produced a proposal for a Large Combustion Plant (LCP) Directive which would have spread German standards of control across the EC. The motivation for the LCP Directive was as much the harmonization of regulatory controls, so as to reduce barriers to trade, as the improvement of environmental standards *per se*.

The proposed LCP Directive was resisted by several member states, led by the UK. When Spain and Portugal joined the EC in 1986, the CEC's original proposal became untenable. A process of bartering began which led to the negotiation of a much watered-down Directive in June 1988.

The UK's resistance to the LCP Directive was partially conditioned by the relative lack of public and political interest in acid rain as an issue. Other underlying reasons included the lack of assertiveness on the part of the pollution control authorities, the absence of new markets for power station equipment in which to test control technologies and tight controls on public expenditure which had a considerable bearing on the investment plans of the nationalized electricity supply industry (ESI). These factors resulted in an official emphasis on the need for conclusive scientific evidence about the effects of acid rain in advance of remedial action.

The UK has now had to accept the need for a national plan for the abatement of SO_2 and NO_x under the terms of the LCP Directive. The time-scale for emissions reductions is however very long, with the final stages of the plan only to be achieved by the year 2003.

Consequently, many of the issues debated in the FR Germany as the GFAVo was being formulated began to surface in the UK only in 1990 as detailed plans for complying with the LCP Directive began to be formulated. Issues, intimately tied to the future of the UK's energy industries, particularly in light of the Government's privatization programme, include the extent of reliance on the FGD technology, the size of future markets for British coal and the appropriateness of using natural gas as a power station fuel.

What this story, briefly told, reveals is the very long time-scales over which large scale international environmental problems such as acid rain must be resolved. From the start of serious international pressure on the part of the Scandinavians in 1972, it took sixteen years before reluctant countries, such as the UK, were brought within the umbrella of any legally binding international agreement. It will be another fifteen years before the undertakings embodied in that international agreement have their full effect.

Within such a long history, social, political and economic issues inevitably play major roles. It is the purpose of this book to unravel these factors as they influenced the course of the acid-rain debate.

Acid rain derives from emissions from many sectors, not just the energy industries which form the main topic of this book. Transport is the major source of nitrogen oxides (NO_x), which became subject to increasingly stringent regulation during the 1980s. Vehicle emission regulations are dealt with only to the extent necessary to provide a reasonably complete picture of acid-rain policy. This is particularly necessary for FR Germany where the clean-up of power stations and passenger cars took place simultaneously. Britain fought against tight vehicle emission controls until 1989, when the EC decided to require the installation of catalytic convertors on all new cars from the early 1990s onward.

The framework for analysis

This book is essentially empirical. The object has been to develop insights into styles and patterns of environmental policy-making in representative conditions entailing differences in energy policy, international perceptions, ongoing technological change, inevitable changes to the terms of competition in affected markets and an uncertain knowledge base. Both the obstacles to better environmental protection and the forces and conditions which help to enhance it are illuminated.

The questions which the book addresses are not, therefore, normative in the sense of prescribing policy. Rather, the aims are to examine the conditions for successful national and international policy-making and to enhance understanding of the energy–environment policy interface through the study of public and corporate decision-making.

The social and political processes relating to the acid-rain problem are studied empirically, both 'top-down' by looking at public policy-making at the international and national levels, and 'bottom-up' by looking at British and German societies in considerable detail. By focusing neither on the science of the acid-rain problem nor on 'ideal' solutions, the role of science in resolving large-scale environmental issues is implicitly questioned. A related issue which arises is that of the relative roles of science and technology in the policy-making process. While science enables an understanding of problems and can indicate the broad nature of possible solutions, it is technology which provides the fundamental capability to respond.

The very divergence of policy responses in the two countries invites a search for broad explanations based on a close examination of the course of events in each country. Emphasis therefore falls on the analysis of contextual factors and forces.

As environmental issues have now become a major element of international diplomacy, a second objective of the book has been to gain a better understanding of the inherent potential for conflict, how and why environmental disputes may arise and how they might be mediated. The analysis reveals that national culture, domestic politics and institutions play vital roles in determining the international positions of states.

While the primary focus of the book is the examination of national policy formation, the importance of international organizations, such as the UN and its constituent programmes and commissions, and supra-national organisations, such as the EC, is acknowledged. The particular role of the EC in helping to mediate the acid-rain issue in Europe is described in some detail. Nevertheless, the scope for action on the part of international bodies is circumscribed by the willingness and fundamental ability of individual nations, given internal social, economic and political constraints, to act. This will undoubtedly become clear as diplomatic efforts to resolve the even larger problem of global climate change gather momentum.

The determination of national policies

There are several broad schools of thought concerning the determination of national environmental policies and attitudes to the environment. One school would emphasize differences in cultural values as the fundamental determinant of national positions. Another would look towards an analysis of economic interest of the major actors participating in policy formation. The prime concern for others would be institutional frameworks and the long- or shorter-term interests of government.[2]

The approach taken here is eclectic. The underlying hypothesis is that all factors need examination and that 'real' policies are the result of many forces. Concerns about the environment are undoubtedly filtered and directed by cultural and social factors. The role of interest groups may become paramount when issues are being debated and specific solutions are selected as being the most appropriate or acceptable. In this process of selection and direction, institutional commitments and capacities also play a major role, especially in competitive conditions.

The linkages between cultural aspects and motivations of interest are extensive and subtle. Senior civil servants and company executives are citizens as well as managers. They will be influenced by their personal views and, sometimes more importantly, by the views of their family and friends. Companies and political parties must take account of the evolving attitudes of their customers or supporters.

Culture is not fixed, but like language, evolves continuously. It is to some extent malleable through education and media influence. Governments and companies can influence public attitudes through advertising, example and the judicious release of information. 'State sponsored environmentalism', through the campaign to save German forests, was an important factor in promoting public concern and support for the German government's policies in the early 1980s. These measures, by themselves, furthered objectives in policy domains other than the environment. Since the British prime minister's 'green' speech to the Royal Society in 1988, there are signs that a similar relationship between government and the public may be developing in the UK.

Cultural differences

'Culture' is a vague term with many dimensions. It can encompass popular culture, as well as the attitudes and styles which characterize corporate enterprises, economic management, political relationships and regulation. While these distinctions are recognized within this book, all of the components apart from national culture are subsumed into the analysis of institutional differences.

The elements of national culture assessed include attitudes to the environment and technology (often expressed in concepts embodied in very different linguistic approaches), views on the relative responsibilities of citizens and the state, susceptibilities to perceived external threats, and different needs for national symbols and socially cohesive national objectives. Attitudes to 'Europe' and the world beyond national boundaries and about the need to make provision for the future are other aspects of the same general cultural theme. Each of these elements is shown to have conditioned the broader social responses to environmental problems in Britain and Germany.

It is recognized, however, that popular culture and attitudes can evolve rapidly and that, in many ways, the environmental debate in the UK is taking on some of the characteristics of that in Germany a decade before.

Institutional frameworks

Considerable attention is devoted to institutional frameworks and their influence on policy-making. Institutions promote some interests and suppress others. They filter perceptions and select solutions which are seen to be legitimate and politically acceptable. The definition of the roles of different actors in the acid-rain debate in each country, and the relationships between them, have had significant impacts on policy choices and developments in each country, and help to explain the divergence in policy responses.

Strong differences in institutional arrangements which affected acid rain are found in the political system, the structure of government and its constituent ministries, and in the relationships between central/federal government and the regions and municipalities. The organization of the energy sector and arrangements for air pollution control also differ significantly between Britain and Germany.

Differences here have their origin in factors which cannot be changed. While Britain has a long and relatively stable political history, the course of events in Germany has been more turbulent. Many of the FR Germany's institutions, including its Constitution, are relatively recent, having been established only after the Second World War. On the other hand, these new institutions co-exist with more traditional tendencies, such as those of industrial concentration and 'corporatism'.

The role of political and economic interests

The fundamental premise underlying much of the analysis in this book is that the creation of the political will to act on a particular issue, or to resist that action, is dependent on the creation of sufficient 'winners' who can derive advantage from a proposed solution.

The main economic interests identified are the different energy supply companies. The electricity supply industry, where acid-rain controls had profound impacts on costs, technology and fuel choice, was at the heart of the debate in both Britain and Germany. The fortunes of the indigenous coal industries were also deeply involved. The issue of the use of domestically produced coal versus alternatives, such as coal imports, the development of nuclear power or the use of natural gas, was at the heart of the economic debate about acid rain.

However, these purely economic interests were superimposed on a set of wider political interests. Acid rain controls became intertwined with macro-economic policy making in both countries, while regional politics became an additional issue in the FR Germany. Also, in Germany, where coalition government is the rule, acid-rain policy became critical to the interests of individual political parties. This was not the case in the UK.

The analysis of political and economic interest is thus a major theme of the book.

Organization of book

The fourteen chapters of the book fall into four parts. Part I, comprising Chapters 1 to 3, establishes the wider context within which British and German acid-rain policy was set. Chapter 2 examines the air pollution issue both from a historical and international regulatory perspective, describing the development of international environmental treaties in general, and those on air pollution in particular. Chapter 3 provides a very brief description of the science of acid rain, attempting to abstract those issues which are most relevant from the point of view of the policy process. This chapter also considers, in more general terms, the role of science, expertise and rationality in the formation of public policy.

Part II comprises the contextual chapters (4 to 9) which describe the relevant differences between Britain and Germany. Broadly, the chapters move from the 'softer', less tangible issues such as differences in culture, to 'harder' questions such as economic management and energy policy.

Cultural differences are described in Chapter 4 with particular emphasis on environmental threats and the different perceptions of the links between environment and society. Having outlined cultural predispositions towards environmental issues, Chapter 5 moves on to examine the emergence of the environmental movement and green politics in both countries. Green politics developed earlier in Germany and has, unlike in the UK, achieved parliamentary significance. While, as far as possible, a directly comparative approach has been used, a degree of emphasis towards German developments has been unavoidable in Chapter 5.

Chapter 6 describes the structure of party politics and the system of government in the two countries, drawing out those features which had the greatest impact on the development of acid-rain policy. The reasons for the greater impact of green politics in Germany are explored. The differences between Germany's federal structure and the much more centralized nature of decision-making in Britain are shown to have been important in determining the course of events.

Economic performance and economic management are the subjects of Chapter 7. This chapter is intended to establish the roots of the very different willingness to pay for environmental controls in the two countries. The two main issues addressed are general wealth and styles of economic management which led to expensive pollution control measures being viewed from very different perspectives.

Chapter 8 examines the structure of the energy sectors and energy policy in the two countries in some detail. There is a particular emphasis on the electricty industry. The electricity supply industry was at the heart of arguments about acid rain, while being the subject, in both countries, of considerable controversy in its own right. Linkages between the coal and electricity industries are described, as are the ways in which environmental controls affected the fundamental economic interests of different players in the energy markets.

Chapter 9, the final chapter in Part II of the book, describes and analyses the institutional frameworks for air-pollution control in Britain and Germany. There are major differences between the two countries in terms of the historical development of their respective systems and the legal and administrative frameworks. These factors led to a different balance of forces when the acid-rain issue came to be debated during the 1980s. Again, the federal nature of the German system, as opposed to centralized decision-making in Britain, had an important bearing on policy developments.

Part III tells the story of acid-rain policy-making in the FR Germany (Chapter 10), the UK (Chapter 11) and the European Community (Chapter 12). While these chapters are essentially narrative in nature, some differences in structure proved necessary because of the nature of the developments described. Policy-making in Germany was the result of a complex convergence of broad cultural factors and political and economic interests. Consequently, a pure narrative approach was less appropriate. However, in Britain, decision-making was much more narrowly based and it has been possible simply to relate the sequence of events. There is also some discrepancy between the German and British chapters in terms of the period of time covered. Major German decisions made in 1982–83 effectively set the framework for the British debate which took place from 1984 onwards.

Since the activities of the European Community are not the focus of the book, Chapter 12 is relatively self-standing. It describes first the development of Community environmental policy and second the development of legislation on acid emissions from power stations and larger industrial plant. The UK and the FR Germany were key actors in the development of the LCP Directive and their roles are examined in some detail. Although vehicle emission controls developed within the EC are a major component of its acid-rain policy, these negotiations are not described in detail.

Part IV of the book (Chapters 13 and 14) comprises the conclusions of the study. Chapter 13 identifies the implications of the acid-rain debate for the energy sector and the conduct of environmental policy in Germany, Britain and the Community. Developments which have followed the events described in Part III of the book are briefly described. In Chapter 14, the wider implications of the case study are drawn out. The conclusions take account of scientific uncertainty, technological change, political salience, economic interests and different institutional frameworks.

No simple model is postulated – one overwhelming conclusion from this book is that environmental decision-making is complex and multi-faceted indeed. It is not possible to detach ideal solutions to any environmental problem from the institutional and cultural context within which the decisions are made.

Notes

1. For example (in English): G.S. Wetstone and A. Rosencranz (1983), *Acid Rain in Europe and North America: National Response to an International Problem*, Environmental Law Institute, Washington DC; S. Elsworth (1984), *Acid Rain*, Pluto Press, London; N. Dudley, M. Barrett, and D. Baldock (1985), *The Acid Rain Controversy*, Earth Resources Research, London; F. Pearce (1987), *Acid Rain*, Penguin, Harmondsworth; C.C. Park (1987), *Acid Rain: Rhetoric and Reality*, Methuen, London; J. McCormick (1989), *Acid Earth: The Global Threat of Acid Pollution* (2nd edn), Earthscan, London

2. H. Maarten and M. Schwarz (1989), Acid Rain and the Cultural Climate, Paper to the congress on *Scientific Controversies and Political Decisions*, Arc-at-Senans, September 1989, stress cultural factors and thus the British understanding of science and its role in policy. D. Vogel (1986), *National Styles of Regulation*, Cornell University Press, places more emphasis on institutional interests and traditions, while theorists of both the Marxian and public-choice schools emphasize economic interest.

2 Atmospheric pollution control and international relations

Introduction

While the primary focus of this book is the development of environmental policy in Britain and Germany and, to a lesser extent, policy formation within the European Community (EC), the wider context of international relations and attempts to forge air pollution control agreements at that level remain important. In particular, acid-rain damage in Scandinavia was the original driving force behind the atmospheric pollution debate in Europe and Norway and Sweden have proved adept at using international institutions to seek redress for their grievances. This chapter considers this wider context.

Consideration is given first to the way in which pollution may be conceptualized within public international law and its implications for negotiations. International remedies for transfrontier pollution (TFP) are then considered, placing emphasis on the importance of the development of treaty regimes. Obstacles and limits to treaty law are discussed, as are the international legal instruments available. Before specifically addressing the development of international law on air pollution, the earlier example of marine pollution is assessed, as important lessons were learned from this experience.

The development of concern about acid rain is then described, showing how first the OECD and, subsequently, the UN Economic Commission for Europe (UNECE) helped to facilitate wider international agreement on transboundary air pollution. The specific treaties and protocols which have been agreed are discussed. The special case of the Treaty of Rome which links EC Member States is given brief consideration, with a fuller discussion postponed until Chapter 12.

The concept of pollution

It is now almost axiomatic that governments are for the environment and against pollution. However, for international relations, the problem of defining pollution is fundamental, as different definitions have implications for the acceptance of international obligations. National definitions may not coincide with international ones, giving rise to problems during both the negotiation and the implementation of treaties.

The problem of meaning can be reduced to two questions. How is pollution to be identified, and how can it be measured? Without agreement on these questions a rational resolution of environmental problems by means of regulation becomes virtually impossible. While many shades of meaning are possible, it is helpful to distinguish between two broad types of definition for the term pollution. The first, pollution as an effect, appeals more to British tradition and to international law. The second, pollution as the simple presence of undesirable substances, is more compatible with the view adopted in FR Germany and, increasingly, within the EC.[1] The former emphasizes the need for evidence of environmental damage and may appeal more to scientists. The latter is more accommodating to the view of engineers responsible for abating pollution and to regulators directly associated with implementation. In practice, a link between the two has to be established. This link cannot be scientific, but must, in the end, be a matter of value and ethics determined at the political level.

Pollution as an effect

First, pollution may be identified with harm, deleterious effects or damage caused by human introduction of materials into the environment, i.e. as the undesirable consequence, rather than the mere presence of pollutants. From this, it follows that undesirable materials present in low concentrations, widely dispersed or transformed by natural processes, may turn out to be quite harmless. This concept of pollution is widely accepted in English usage and corresponds to the precise dictionary definition of the word pollution.

This definition is the traditional basis of international marine pollution treaties, particularly the detailed technical clauses and annexes as opposed to the lofty preambles. It requires the demonstration of unacceptable damage before precise standards can be negotiated. There has been a long, and eventually successful, attempt within the UN to weaken this definition by defining pollution as materials *likely* to have undesirable effects rather than materials *proven* to cause damage. The inclusion of the world 'likely' clearly points to the development of a more precautionary approach.

Defined as an environmental effect, pollution becomes a concept with which scientists and lawyers can live and prosper. Pollution requiring scientific proof of causality creates the need for evidence which will stand up to the tests of scientific rationality and legal cross-examination. It leads to an operational definition of pollution which is not easily quantified, except through comparatively sophisticated concepts such as dose-response relationships, the absorptive capacity of environmental media and critical loads on eco-systems. In principle, pollution in terms of effects may also be expressed in monetary terms, but only after cause and effect relationships have been clearly established and the controversial process of assigning monetary values to environmental assets has been undertaken. However, even if this were achievable, consensus on abatement measures would by no means be assured.

The distribution of the costs and benefits of the abatement of pollution defined as unwanted damage is typically very uneven. When this factor enters intergovernmental negotiations, it tends to lead either to conflict couched in scientific terms, or, if solutions are genuinely sought without resort to political pressure and coercion, to economically motivated bargaining between polluters and their victims. Pollution defined in this way also tends to encourage a wait and see approach, i.e. the postponement of abatement efforts until scientific evidence becomes extremely clear. This tends to favour the polluter and exporter of pollutants, especially if the legal regime places the burden of proof on the victim rather than requiring proof of harmlessness from the polluter. Given the facts of the acid-rain controversy, it should not come as a surprise that Britain has, until very recently, preferred the definition of pollution as environmental damage.

Pollution as an undesirable material

Pollution may also be identified simply with the emitted materials themselves, e.g. acidic gases, or with the quantity which is discharged. In principle, pollution defined in this way is much more easily measured using physical rather than biological parameters, such as the total quantity of pollutants discharged over time, or their concentrations in the waste streams.

There are limits to the value of such a definition. For example, in 1980, the base year for emission reductions required under the Helsinki Protocol on SO_2 and under the EC's Large Combustion Plant Directive, the UK emitted 3.8 m tonnes of SO_2 and the FR Germany 2.2 m tonnes. This apparently makes Britain the larger polluter. However, a large proportion of the UK's SO_2 is absorbed in the ocean where it is thought to do no harm. At the international level there may be no alternative to allocating responsibility in this way.

The regulation of pollution is enormously simplified by this definition because it ignores the complex relationships between the quantity of a material introduced into the environment and subsequent harmful or undesirable effects. It is a deceptively common-sense definition which is unacceptable to both science and law – unless governments agree to it. The German language itself, as explored in Chapter 4, encourages this definition by describing pollutants as *Schadstoffe* (harmful substances) and pollution as *Verschmutzung* (making dirty).

Pragmatic approaches

International law can thus do little about pollution until there is: (1) agreement on both the evidence for environmental damage; and (2) a consensus that this damage, nuisance or interference is no longer acceptable to all the parties. A polluter may of course reduce pollution exports voluntarily and unilaterally. This may be done out of self-interest, or because the polluter is prepared to clean up in the common interest. Doing the latter noble deed would mean states implementing a fairly recent international legal principle agreed under OECD auspices (see below). However, if they do not, then the effort to abate transboundary pollution must return to the international political arena. Once responsibility for damage is agreed, international institutions must address the cost-benefit balance for the participants in the decision-making process.

Legal remedies for transboundary pollution

In principle, three legal solutions to transfrontier pollution (TFP) problems are possible: (1) liability regimes; (2) equal right of access of foreign pollution victims in a polluting country's national decision-making and litigation procedures; and (3) international treaties incorporating general abatement requirements, or, possibly, precise standards and abatement timetables for each state. No single option can solve all environmental problems. In addition, each option requires the clear identification of pollution sources and the consequent environmental damage.[2]

Liability

Chronic, cumulative environmental damage, such as that caused by acid deposition, cannot be resolved through liability regimes because the associated uncertainties about who is at fault and the evaluation of damage remain too great. There are no signs of such a system developing for air pollution. To be effective, it would require all the countries concerned to set up insurance funds and accept some international judicial or arbitration body to allocate blame and assess compensation.

However, such a system is appropriate for catastrophic environmental damage which follows, for example, from accidental oil spillage or nuclear disasters such as Chernobyl. Liability regimes can, through insurance mechanisms, provide important incentives to take precautionary measures against incidents taking place. These incentives are particularly strong for insurers who underwrite against accidents. Establishing liability for air pollution damage remains, however, very elusive even at the national level. German forest owners, for example, have failed to obtain financial compensation for alleged air pollution damage to their trees through the courts.

Equal access

The promotion of rights of equal access to environmental policy-making and judicial processes for foreign parties has made little progress. British views were not formally heard in the FR Germany during the consultation stages in its domestic policy-making process. While the UK Parliamentary Select Committees heard German and Scandinavian opinions when they enquired into the acid-rain issue, this had little legal significance. Within the EC it can, of course, be argued that bargaining and debates in Brussels, in the Commission, the Council of Ministers and, increasingly, in the European Parliament render equal access at the national level redundant. However, by allowing stronger mutual influences on policy-making processes, such access might assist environmental policy by lowering political costs and encouraging a more positive, informed and democratic policy-making process.

Intergovernmental treaties

Negotiations leading to an international contract, i.e. a treaty, convention or protocol, have been the primary method by which countries have tried to establish intergovernmental regimes to alleviate pollution. While these regimes are legally binding, they are essentially unenforceable at the international level and remain, in essence, voluntary. Solutions cannot be imposed. Enforcement takes place only at the national level, although information and reporting obligations improve the verifiability of such agreements.

Before treaties can be negotiated a great deal of preparatory diplomatic work is essential. For TFP, the OECD played a major role in this respect during the 1970s.[3]

Since treaty regimes benefit from the ceding of sovereignty by their members, a distinction must be made between the more purely intergovernmental regimes developed under UN bodies such as the UNECE and the Environment Programme (UNEP), and that developed within the EC.

The Treaty of Rome

The Treaty of Rome which links the EC Member States is not a major topic of this chapter. However, the particular features which distinguish it from other types of international treaty are noted here. The Treaty of Rome has a special status because a degree of national sovereignty is ceded to a supranational body and contracting states submit themselves to the legal decisions of the European Court.

The EC regime is much the most demanding in international law and is based on the ideas of fair competition, the removal of trade barriers and, latterly, since the 1987 Single European Act, the protection of a common European environment.[4] The justification for harmonized environmental standards is as much commercial and economic as it is environmental. Britain and Germany have frequently found themselves on the opposite ends of the negotiating tables because of their different economic interests. Once the commitment to harmonization is accepted, it follows from the pattern of transboundary data noted in Chapter 1 that Central European countries, such as Germany, have a continuing incentive to move towards greater stringency of common standards simply to solve their own air-quality problems.

EC regulations are directly binding on individuals and companies within each Member State. However, EC environmental policy has generally been pursued through the issuing of Directives which, like traditional international protocols, are binding only on nations and do not apply directly to individual citizens or corporate bodies. The interests of individual EC Member States may be overridden by the common interest. Some forms of sanction are available to the EC and, since the 1987 Single European Act, unanimity in decision-making is no longer required in all circumstances, i.e. individual states can be outvoted.

Broadly speaking, the nature of the EC regime facilitates agreement among its members, although particular topics of major concern to individual States may prove no more tractable than under any other treaty regime. Yet, the chances of reaching agreement based on the lowest common denominator are enhanced, even if it takes years of political pressure and horse trading. However, the Council of Ministers still exercises decisive power within the EC, often giving it the flavour of little more than an intergovernmental forum.

It may even be the case that the negotiation of certain global problems in which the EC participates as a distinct legal entity are hindered by the length of time which it takes EC Member States to agree a common position. It has been suggested, for example, that this was the case for the Montreal Protocol on CFC emissions.[5]

Obstacles to the negotiation of treaties

Sources of air pollution are primarily land-based and are likely to be the subject of national law prior to the negotiation of treaties which may govern TFP. In international law it is therefore governments representing individual states, and not an industry, a firm or an individual, which are considered to be polluters or pollution victims. This means that any state which enters into an international commitment to control pollution, whatever this may mean in practice, must ensure that it does have jurisdiction over domestic sources of pollution. If a treaty is to be effective, governments must have the capacity to implement their obligations to the satisfaction of international partners.

In the process of adjusting national and international law, or adding international obligations to existing national ones, great diplomatic and legislative efforts are usually required. This tends to take years, or even decades, rather than months. The process of implementing environmental obligations is also a continuous process requiring repeated adjustment to advancing knowledge and changing priorities. Many factors have influenced the development of intergovernmental responses to transboundary air pollution. Broadly, these stem from the land-based nature of the pollution sources.

Until a decade or so ago the abatement of air pollution remained an entirely national task. This left countries free to adopt control measures of variable effectiveness on the basis of legal or administrative mechanisms of their own choice. States generally ignored transboundary exports of pollutants, either because they were unaware of them, or because the atmosphere (like the ocean), was considered a natural and tolerant sink for wastes. As major industrialized and densely populated countries, FR Germany and Britain developed national air pollution control regimes based on their own particular experiences, legal and administrative institutions and patterns of industrialization (see Chapter 9). The ease with which these two countries could respond to international demands for more stringent regulation necessarily had an impact on international discussions, particularly within the EC.

The purely national nature of the responsibility for air pollution also meant that any attempt by other states or international bodies to alter these national regimes immediately brings into play questions of sovereignty and the competence of a national bureaucracy. Once allegations are made against countries or their subjects, powerful defensive impulses from those identified as polluters may be expected.

It is also difficult and time-consuming to persuade governments busy with domestic affairs to negotiate, let alone implement, treaty regimes which they perceive to be against their own interest. Any law, especially an international one requiring expensive domestic regulation, implementation and enforcement, is likely to be resisted, especially by governments with limited resources and other priorities. There are no other mechanisms other than international power politics to alter this state of affairs.

Once it has been accepted that a treaty is desirable, there remains the all-important question of reaching agreement on what constitutes pollution at the international level and how it can be measured. Implementation, enforcement and monitoring of agreed measures are subsequent tasks upon which the effectiveness of the agreement will of course depend. The best policies and laws are of little use if they do not work, are not obeyed or do not have the desired effect.

Finally, as implementation progresses from standard setting to enforcement, the interests and powers of top national decision-makers tend to become weaker and the responsibilities of industry and individuals to comply with environmental regulation grow. If regulation does not, in the end, become accepted by the organizations which are being regulated, the costs of enforcement may make the entire regime unenforceable.

Acid rain and other major environmental problems are not, therefore, matters which readily attract government attention and interest. Problems will be ignored as long as possible, unless or until major political interests are at stake. This may happen much earlier for some countries than others.

Limits to treaty law

Since intergovernmental treaties are entirely voluntary affairs, states can withdraw from them. Treaties gain their bite more from politics than international judicial institutions. Environmental treaties, being a rather new international legal species, first require agreement on general principles about objectives, fairness and conflict resolution.[6]

Under conventional international law, legal competence may simply be allocated to the individual countries. This means that both the making of policy, including the setting of longer term objectives and precise regulations, and their implementation and enforcement, remain entirely national responsibilities. However, these tasks may also be divided in varying degrees between national and intergovernmental authorities. The most customary division of labour for dealing with transboundary problems has been the intergovernmental definition of principles embodying objectives, followed by the adoption of general treaty objectives which are later given precise meaning through protocols. These are added to the original treaty and set international standards, but leave implementation and enforcement to national authorities.

International norms, described in more detail below, may include environmental quality standards which states are left to measure and achieve as best they can, precise technical emission standards for individual sources measured by agreed testing procedures, or limits on total national emissions for specific pollutants accompanied by timetables for their adoption and reporting obligation to international bodies. Technology transfer and direct payments between contracting states may also be included. If implementation and enforcement are left entirely to national authorities, it is they who, in the end, will determine whether a treaty is effective or not.

Without the adoption of commonly agreed regulatory instruments by states, emissions with transboundary environmental impacts remain subject to vague legal principles, such as good neighbourliness and reasonableness in the use of common resources.[7] These principles have been of little help to countries which consider themselves adversely affected by foreign emissions. Such countries have therefore adopted the only alternative available to them, the use of political pressure and persuasion exercised through existing organizations such as the OECD or the UN and its daughter organizations. An alternative is to attempt to set up new bodies which are allocated the task of developing environmental regulatory regimes, either by treaty or informal agreement.

New institutions, especially intergovernmental ones, are costly and are not easily set up. More commonly, existing institutions attempt to increase their competence by absorbing new issues and responsibilities. An example is the way UNEP has assumed responsibility for coordinating international efforts on global warming. Such institutions, while not making decisions themselves (this remains the duty of governments), gather information, arrange agendas, solicit advice, make proposals and may even bring considerable pressure to bear on negotiating parties.

The sharing out of competences and pollution abatement tasks between national and intergovernmental bodies also raises questions of motivation and political power.[8] As more choices and decisions are removed from national and local levels and transferred upwards to international bodies and officials without a direct stake in their implementation, the incentive to act at the national level may decline as power and responsibility are lost. Former decision-making roles are reduced to mere implementation and enforcement. Resistance to inter-governmental regulation is therefore often very strong for political as well as economic reasons, and powerful special arguments are generally needed for the justification of international regulation. This is particularly true for regulation which is mandatory rather than, as is much more common, merely advisory, as with World Health Organisation standards. All of these points are illustrated by the slow development of the acid-rain treaty regime described below.

Treaty instruments

International treaties of an environmental nature, for example the LRTAP Convention, are frequently worded in very general terms. However, the treaties themselves may act as stepping stones to more specific agreements between contracting states which apply quantitative measures to the limitation of pollution. Protocols to the original treaty may be the vehicle for reaching such quantitative agreements. It is not necessary for all the original parties to a treaty to sign up for any particular protocol, but an agreed number must have ratified before a protocol becomes binding. In the mean time, those who have signed are bound not to take any action which would prejudice achievement of the common goal.

National bubbles and other instruments

The most common quantitative instrument used in protocols for air pollution control has been the national 'bubble' – a limit on the total annual quantity of a pollutant produced within national boundaries. This may take the form of a standstill in emissions by a given date, referred to a specified base year or a fixed percentage reduction. At the international level, this is the simplest type of quantitative agreement to negotiate. However, scientific and technical uncertainties mean that states have some flexibility in establishing methodologies for estimating emissions and defining base lines. While a number of national atmospheric emission bubbles have already been negotiated under different treaty regimes, no agreement has as yet become binding. International agreements on SO_2 and NO_x will start to bite from 1993 onwards and it will then be possible to reach some judgement about their effectiveness.

The emission of acid gases arises from two classes of source – stationary sources such as power stations and industrial plant and mobile sources such as motor vehicles. The two major types of source have been subject to quite different legal and administrative systems of regulation at the national level. In addition, emissions from each type of source may be influenced by product standards applying to the characteristics of fuels.

If any intergovernmental agreement is to go beyond the setting of national bubbles, these two types of source need to be considered separately. In practice, the distinction has proved to be particularly important within the context of the Treaty of Rome. Prior to the acid-rain debate, only mobile sources were governed by intergovernmental technical norms, and even then only to a very limited extent, whereas power stations remained an entirely national responsibility.

Depending on the country involved, 40–60 per cent of man-made nitrogen oxides come from mobile pollution sources. Pollution from these may move directly from one national territory to another as part of air masses, or indirectly as emission sources in the form of traded goods or even by being driven there by foreign nationals. Questions of trade barriers and the regulation of foreign assets can therefore become a subject for environmental negotiation. Britain and Germany trade motor vehicles with each other, with Germany a substantial net exporter of private cars. If emissions standards differ, as they still do, then different products must be produced to meet the requirements of different markets. The international regulation of motor vehicle exhaust emissions therefore has enormous industrial and commercial implications.[9]

Emission standards for vehicles have been the subject of UNECE negotiations since the late 1960s, and have been regulated by the EC since 1983. While uniformity of standards is attractive to the motor vehicle industry, the stringency of measures, and thus the technology needed and costs involved, were major areas of conflict until 1989. Negotiations have proved to be at least as difficult as those for power stations and other stationary sources which are the main focus of this book.

The European car regime now involves a distinct set of 'homologation' procedures for what is, in fact, the world's major internationally traded good. The manufacturers of traded pollution sources tend to have a strong interest in the establishment of uniform or harmonised regulations, or, at least, the mutual recognition of emission standards.

Until 1988, there were no international agreements applying specifically to stationary sources. Power stations and other large emitters are now directly regulated under the EC regime to be described in Chapter 13. Removing SO_2 and NO_x from flue gas streams has important implications for technology choice, the economic lifetime of plants and fuel choice. The Cost implications of international agreements have bedevilled intergovernmental negotiations, as later chapters of this book will show.

Given the nature of the Treaty of Rome, fuel quality standards were among the first types of environmental regulation to be developed in the EC. In 1976, limits were established for the sulphur contents of gas and diesel oil, whether burned in stationary sources or in motor vehicles. A later attempt to introduce similar standards for the sulphur content of fuel oil foundered because it proved impossible to secure political agreement.

Lessons from marine pollution control

On purely scientific grounds, international priority might reasonably have been given to air pollution problems during the 1970s, given the evidence presented to the 1972 Stockholm Conference. For the purpose of treaty-making and institution building, however, the global political system picked up marine pollution issues and, for over a decade, the international agenda remained devoted to the building of an international regime for marine pollution control.[10]

This proved to be both less demanding and politically more beneficial to individual countries because of the threatened revision of maritime law. This was demanded by a large number of countries in the early 1970s as part of the 'Common Heritage' and 'New Economic Order' movements. The revision of the law of the sea was not completed until 1982. A number of global and regional treaties on marine pollution were negotiated both within the UN system and separately by smaller groups of governments.[11] Air pollution control was to benefit from this experience.

With its largely laissez-faire regime, traditional maritime law favoured countries with large navies, research vessels and commercial fleets. The freedom of the high sea, a highly permissive transit regime for commercial and military vessels, coupled with weak national powers over seabed and water column resources, were both challenged by a concerted global attempt to prevent the militarization of the ocean floor as well as the economic exploitation of the ocean and the seabed by those most capable of doing so. Negotiations to revise this regime began in the late 1960s with attempts to change the scope of international law, by setting up more intergovernmental organization and by giving more competence to national authorities.

In particular, serious disagreements arose between coastal states and traditional maritime powers over the regulation of transport and rights to economic resources in and under the ocean. In each case, it was marine pollution, and hence the environment, which was used as an argument in justification of: (1) setting up a stronger international regime in order to protect the global commons; or (2) strengthening national powers, so that more equity and self-interested protection of the sea would be achieved.

Coastal states, and especially Canada, argued that nations could better protect their resources and transport routes from the effects of marine pollution if they were given more rights over the marine environment. This gave marine pollution, in the public eye at the time because of several major oil spills, an extraordinarily high political profile. To remove marine pollution as a substantive issue from these primarily political battles, the major polluting countries set up anti-pollution treaty regimes under which marine environmental science and law for the oceans began to develop. Experience was also gained in institution building which subsequently served as models, both positive and negative, for air pollution.

The regulatory regime developed for the ocean environment, weak as it is at the global level, is, however, very strongly based on science.[12] It therefore remains in the hands of those countries with the greatest scientific capacity and resources.

The emergence of acid rain

There were no comparable jurisdictional threats to the international legal regime covering the atmosphere. Air space was already largely under national control and only the turbulent atmosphere itself knew no boundaries. Also alerted to the threat of acid deposition, polluting countries continued to wait for proof of damage until the end of the 1970s.

Sweden and Norway, in particular, had not remained idle, in the meantime, but made considerable use of international organizations to advance their case for stricter air-pollution controls. Since neither is a member of the EC, these efforts were focused first on the OECD and subsequently the UNECE. Scandinavian activity and the achievements of the OECD and the UNECE are described below.

Swedish scientists have been monitoring precipitation patterns in their country since the 1940s (see Chapter 3).[13] By 1968, the increasing acidity of rain over a period of time had been linked to sulphur emissions in the UK and Continental Europe. In 1972, the UN Conference on the Human Environment attended by most countries (but not the USSR because of the still unresolved problem of the recognition of East Germany) took place in Stockholm at the initiative of Norway and Sweden. Sweden presented evidence at the conference on the extent of the long-range transport of pollutants and the damage which was being caused to soils and lakes.

This point of view was initially met with a great deal of scepticism by the world scientific community. However, over the last 15 years, the Scandinavian position has become the conventional scientific wisdom. Britain turned out to be the major sceptic and was not converted until the mid-1980s, although in 1972 the newly formed UK Department of the Environment had acknowledged that:

> UK policy has been to seek natural dispersal and dilution by means of high chimneys. This ... has worked well, but must be regarded as a palliative rather than a cure. We do not know a great deal about what happens to sulphur dioxide emitted from tall chimneys. ... If this (increased acidity in Scandinavia) proves to be the case, and if that increase is harmful ... international agreement is ... desirable, both to control the distribution of low sulphur fuels, and to ensure that the emission of sulphur dioxide from high chimneys in one country does not cause increased pollution elsewhere.[14]

This response clearly laid down the conditions under which Britain would be prepared to do something about her sulphur emissions. The response from the FR Germany was, at the time, very similar.

In Stockholm, all the nations present, including the UK, accepted Principle 21 of the Conference Declaration which states that:

> States have ... the responsibility to ensure that activities within their jurisdiction or control do not cause damage to the environment of other states or of areas beyond the limits of national jurisidiction.[15]

The acceptance of this important principle owed a great deal to the strength of Scandinavian (and Canadian) concerns about both transboundary air pollution and marine pollution.

The role of the OECD

Little concrete political progress was made at the international level with regard to air pollution in the years immediately following 1972 as the world coped with the energy crisis and a major recession, and as international environmental diplomacy focused on the problems of marine pollution. However, the attention given to air pollution survived the 1970s, but at the regional rather than the global level. While the OECD took on the problem of transfrontier air pollution on behalf of the developed nations, there was no comparable effort in the centrally planned economies. In the early 1970s, the OECD set up its Environment Directorate which acted as a source of information for industrialized countries. At the time, acid rain was largely seen as a problem for the Northern hemisphere, long-distance transport of air pollutants being poorly understood and assumed to be regional rather than global.

The principal forms of OECD activity during the 1970s were research, publicity and the development of policy principles and recommendations issued by the Ministerial level OECD Council. However, the lack of any legal mechanisms for implementing OECD policies and principles undermines their operational significance, reducing them to an even more voluntarist form of control than that exercised under international treaties. Nevertheless, the longer-term impact of such advisory and information creating bodies should not be underestimated. They work at the political rather than the legal level. As long as environmental policy-making remains a primarily political and techno-cratic rather than legal process, the influence of such bodies, though often hidden, will remain important.

In 1972, the OECD initiated a 'Program to Measure the Long-Range Transport of Air Pollutants' based on 76 measuring stations in 11 European countries reporting to the Norwegian Institute for Air Research. This EMEP (European Monitoring and Evaluation Programme) network was later taken over by the UNECE. By 1988, it contained 90 monitoring stations in 24 countries and served as an information base for the LRTAP Convention. The early OECD work created initial estimates of sulphur deposition patterns throughout Europe and, although much uncertainty remained, the contributions made by each countries. The results tended to confirm the view that foreign sources were major contributors to acid rain in Scandinavia. EMEP remains the only international monitoring network linked to a treaty.

In 1974, the OECD Council adopted a set of 'Principles Concerning Transfrontier Pollution' which included: 'non-discrimination' – i.e. trans-boundary pollution should be controlled as strictly as that remaining within a country's borders (a rather milder and more realistic formulation than the Stockholm Principle cited above); 'equal access', already discussed; and 'notification and consultation'. This latter principle allowed for mechanisms by which countries should inform international partners in advance about plans which might be expected to increase transboundary air pollution. The need for formulating such a principle is evidence of the low priority which was then attached to the problem of transfrontier pollution.

In the same year, the OECD Council also adopted the 'polluter pays principle' (PPP), defined as meaning that pollution control costs should be borne fully by emitters (without subsidy) so that the price of goods would fully reflect any pollution abatement costs incurred as a result of their production. The PPP is informed, as one would expect from a body such as the OECD, by concerns about economic growth and free trade. This perspective is also important in the context of EC policy. The implication is that environmental standards should be uniform across different countries. However, this often conflicts with a more scientific analysis which might conclude that different eco-systems need different degrees of protection. In addition, different societies may demand different standards of environmental protection. Implementation of principles such as the PPP is therefore limited to certain countries and its interpretation tends to vary.

In the early 1980s, the OECD sponsored cost-benefit analysis work on sulphur dioxide control.[16] However, by this time, it was beginning to reduce its effort on transboundary air pollution, the UK and the FR Germany in fact voting to disband the programme. However, work on the costs of SO_2 and NO_x control technologies is continuing.

It became clear from OECD work that transboundary air pollution in Europe could not be covered comprehensively without securing the involvement of the

Soviet Union and Eastern Europe. The OECD was not, therefore, the most obvious forum for tackling this problem.

The UN Economic Commission for Europe and the LRTAP Convention

By the end of the 1970s, the main focus for broadly based international pressures to reduce emissions had switched to the UNECE, which includes 34 countries from Western Europe, Eastern Europe and North America. Sweden and Norway again led the effort to place transboundary air pollution on the organization's agenda. The opportunity came when President Brezhnev of the Soviet Union suggested in a 1975 speech that the environment was one of three important issues on which East and West shared a common problem (the other two being energy and transport). The Brezhnev speech was designed to defuse Western pressure on the Soviet Union's human rights record by diverting attention to other issues covered by the 1975 Helsinki Agreement on Security and Cooperation in Europe. The environment was a topic which might be expected to divide Western partners since the record of certain countries, notably the US, the UK and FR Germany, was arguably deficient. The theme of environmental protection as a tool of East–West diplomacy re-emerged during the 1980s.

The Scandinavian countries took this opportunity to press their claims within the framework of the UNECE. The result, after two years of negotiation, was the 1979 LRTAP Convention signed in Geneva.[17] The resulting treaty was a compromise, lacking stringent emissions reduction objectives, but it has been an important symbol of environmental progress. Swedish and Norwegian delegates had proposed strict standstill clauses (preventing SO_2 emissions from increasing) and rollback clauses (specifying fixed percentage reductions for aggregate national SO_2 emissions) but this proved not to be possible given opposition from the US, the UK and West Germany.

Instead, the Convention members promised to 'endeavour to limit and, as far as possible, gradually reduce and prevent air pollution' and to use the 'best available control technology economically feasible'. An Executive Body (EB) was set up by the Convention to allow consultation between countries and to initiate relevant research. This has incorporated EMEP, previously supported by the OECD, now the most authoritative source of information on transboundary air pollution. The LRTAP Convention has been ratified by 31 of its 34 signatories.

A complication in securing agreement on the LRTAP Convention was that the EC, in addition to its individual Member States, is a separate member of UNECE. For a while, the Eastern bloc countries refused to recognize it as a legitimate negotiating body, while the EC itself found it difficult to arrive at a consensus among its Member States.

The Scandinavian countries were not satisfied with the LRTAP Convention. On the tenth anniversary of the 1972 Stockholm conference, Sweden called yet another major conference to remind nations of the acidification of her lakes and streams. The 'Acidification of the Environment' conference renewed the pressure on countries like the UK, Germany and the US which were still thought to oppose international acid-rain controls.

Specific conclusions emerged from a series of expert meetings conducted during the course of this conference which summarized the contemporary state of scientific knowledge about acidification. These conclusions pointed towards the need for major emissions reductions. The final declaration contained the statement that 'deterioration of soil and water will continue and may increase unless additional control measures are implemented and existing control policies are strengthened'.[18] Consequently, the Scandinavian countries made a specific proposal that countries should undertake to reduce SO_2 emissions 30 per cent below 1980 levels by the year 1993.

The 30 per cent proposal failed to win the support of the UK, the US, France or the Eastern bloc countries. However, to the surprise of the international community, FR Germany supported the proposal. The German conversion at the international level was the consequence of a complex set of internal pressures which are described in later chapters of this book.

The LRTAP Protocols

Although not accepted at the 1982 Stockholm Conference, the idea of a 30 per cent SO_2 reduction was subsequently advanced by the Scandinavian countries and the FR Germany. The proposal was taken to the EB of the LRTAP Convention in June 1983, where it received additional support from Canada, Austria and Switzerland. In March 1984, an informal group of ten UNECE countries met in Ottowa to initiate the '30% Club'.[19]

The support of the Eastern Bloc countries came only a few months later. Against the background of the missile deployment crisis of that time, the Soviet Union, keen to maintain some avenue of positive communication with the NATO countries, encouraged the FR Germany to host an international conference. At the 'Multilateral Conference on the Causes and Prevention of Damage to Forests and Waters by Air Pollution in Europe' held in Munich in June 1984, the Eastern bloc countries (with the exception of Poland) signalled that they would be prepared to reduce 'transboundary fluxes' of SO_2 by 30 per cent between 1980 and 1993. This succeeded in isolating the US and the UK within the international community, while enhancing the image of the Soviet Union in Western Europe, notably in the FR Germany.[20]

The British position at the Munich meeting, was determined, as will be described in Chapter 11, at the prime ministerial level, and was an embarrassment for UK ministers and senior officials. They were empowered, however, to promise a 30 per cent reduction in British SO_2 emissions by the end of the century.

Against this background of support, the 30 per cent proposal was adopted as an official protocol (the Helsinki Protocol) to the LRTAP Convention in July 1985. Then 21 countries committed themselves to reducing SO_2 emissions (or, in the case of the Eastern bloc countries, transboundary flows of pollution) by 30 per cent by 1993. The '30% Club' came into force in September 1987 after 16 countries had ratified the Helsinki Protocol.

Domestic and international pressures on the UK to sign the Helsinki Protocol have been considerable. The UK Government has consistently resisted these pressures in spite of the fact that, on the basis of economic forecasts in the early 1980s, it might have been perceived that very little extra in the way of pollution control would have allowed the UK to meet the requirements.[21] More rapid growth in electricity demand than originally projected has, by now, put the 1993 target well beyond the reach of the UK.

In addition to an SO_2 protocol, a NO_x protocol was agreed under UNECE auspices in November 1988. This requires a freeze in NO_x emissions at the 1987 level from the year 1994 onwards. Britain, as well as the FR Germany, found itself in a position to sign this Sofia Protocol, which further requires that: limit values defining state-of-the-art NO_x abatement technology for new stationary plant and vehicles be incorporated in national standards; lead-free petrol be introduced within two years of ratification; and discussions on a second-stage reduction should be initiated within six months of ratification.

On the eve of the agreement of the Sofia Protocol, FR Germany and eleven other European countries made a declaration that they would reduce their NO_x emissions by 30 per cent by 1998, taking as the baseline any chosen year between 1980 and 1985. Spain, Portugal, Greece, Luxembourg and Ireland were other EC countries which joined the UK in not supporting this declaration. Thus, the stage was set for further political pressure within the UNECE framework.

Further developments within the UNECE

The UNECE regime has encouraged national measures, set goals and collected information, but it has been weak on reaching agreement on mandatory measures. It cannot strike agreements except at the lowest common denominator standards of abatement and, consequently, one of its main achievements has been the creation of a common information and science base.

As a result of the highly scientific nature of the workshops which underpin political activity within the UNECE, more sophisticated, science-based concepts for emission control are being developed. In particular, the concept of critical loads for pollution burdens on eco-systems (measured in terms of ambient pollution concentrations or the mass of pollutant deposited annually per hectare) has been developed within the framework of UNECE scientific workshops. A critical load is defined as a 'quantitative estimate of an exposure to one or more pollutants below which significant harmful effects on specified sensitive elements of the environment do not occur according to present knowledge'.[22]

Given the tendency to conceptualize pollution as the simple presence of harmful substances in the atmosphere, the critical load concept is unappealing in the FR Germany. However, the UK, concerned with scientific evidence of actual environmental damage, has been more sympathetic. Nevertheless, the reductions in SO_2 and NO_x emissions implied by the direct application of the critical load concept are very high (of the order of 90 per cent) when compared to reduction targets built into existing international agreements. Consequently, the UK is now giving support to the concept of the 'target load' which is, essentially, a critical load moderated by considerations of economic and political feasibility.

To implement the critical or target load concepts requires analytical tools which can trace the impact of emissions all over Europe on particular eco-systems. To carry out this analysis, the LRTAP EB has adopted the Regional Acidification Information and Simulation (RAINS) model, developed by the East–West supported International Institute for Applied Systems Analysis (IIASA), as its chief tool. Reducing pollution loads for particular eco-systems may be achieved by different national distributions of emissions reductions with very different implications for both the absolute level and distribution of costs. A model such as RAINS is essential if the implications of meeting either critical or target loads are to be assessed.

The challenges posed for the conduct of international diplomacy by the integration of the target load concept with such sophisticated computer modelling techniques in the face of influential sceptics such as the FR Germany, may yet prove insurmountable. The greater achievement of these scientific efforts may well be to increase the pressure for international agreements of a more traditional kind, with target loads playing the role of the scaffolding required to construct the edifice of agreement.

Notes

1. S. Boehmer-Christiansen (1990), Emerging principles of international environmental protection, *The Environmentalist*, Vol. 10, No. 2, pp. 95–113.
2. OECD (1974), *Problems in Transfrontier Pollution*, Paris; and OECD (1978), *Legal Aspects of Transfrontier Pollution*, Paris.
3. Examples are the control of radio-nucleides discharged into the deep ocean and the control of PCBs and heavy metals.
4. Directorate-General for Environment, Consumer Protection and Nuclear Safety (1988), *The State of the Environment in the European Community*, CEC, Brussels.
5. E. Rehbinder (1989), US environmental policy: lessons for Europe?, *International Environmental Affairs*, Vol. 1, No. 1, Winter 1989.
6. S. Boehmer-Christiansen (1990), Emerging principles of international environmental protection, *The Environmentalist*.
7. These principles were agreed in the 1972 Stockholm Declaration on the Human Environment. They represent 'soft law' in that they are not binding but have political consequences and they may underpin future treaty developments.
8. See for example S. Andresen and W. Ostreng (1989), *International Resource Management*, Belhaven, London.
9. S. Boehmer-Christiansen (1990), The regulation of vehicle emission in Europe, *Energy and Environment*, Vol. 1, No. 1, pp. 1–25.
10. S. Boehmer-Christiansen (1981), *Limits to the International Control of Marine Pollution*, unpublished DPhil thesis, University of Sussex.
11. S. Boehmer-Christiansen (1981), Marine pollution control: UNCLOS III as the partial codification of international practice, *Environmental Policy and Law*, Vol. 7, No. 2, Elsevier, Lausanne.
12. S. Boehmer-Christiansen (1982), The scientific basis of marine pollution control, *Marine Policy*, Vol. 6, No. 1.
13. Much of this section is derived from G.S. Wetstone and A. Rosencranz (1983), *Acid Rain in Europe and North America: National Responses to an International Problem*, Environmental Law Institute, Washington DC.

14. Department of the Environment (1972), *Pollution: Nuisance or Nemesis?*, HMSO, London, p. 27.
15. Reported in Wetstone and Rosencranz, op. cit., p. 134.
16. OECD (1981), *The Costs and Benefits of Sulphur Dioxide Control*, OECD, Paris.
17. *Convention on Long Range Transboundary Air Pollution* (1979), Geneva, 13–16 November 1979, Cmnd 7885, HMSO, London, May 1980.
18. Reported in Wetstone and Rosencranz, op. cit., p. 148.
19. *ENDS Report 110*, March 1984, p. 22.
20. These developments are described in L. Bjorkbom (1988), Resolution of environmental problems: the use of diplomacy, in (ed. J.E. Carroll) *International Environmental Diplomacy*, Cambridge University Press (1988).
21. C. Davies *et al.*, Control and remedial strategies, in (ed. K. Mellanby) *Air Pollution, Acid Rain and the Environment*, Watt Committee Report No. 18, Elsevier, London.
22. J. Mulder *et al.* (1990), 'Effects of air pollutants on man-managed and natural ecosystems' in Stockholm Environment Institute, *Air Pollution in Europe: Environmental Effects, Control Strategies and Policy Options*, Stockholm.

3 Science, technology and environmental policy making

Introduction

As awareness of environmental problems grows, and as the corresponding scientific effort both enhances our understanding and reveals the extent of our ignorance, the discovery of new issues appears to continue unabated. While, at the political level, acid rain was one of the most significant environmental problems of the 1980s, the problem of global climate change seems likely to dominate policy-making during the 1990s and, perhaps, beyond. At the same time, both the scale and the scope of new problems is increasing.

Environmental problems are becoming more intractable as a result of their scale. Whereas, for example, the impacts of air pollution were once believed to be significant only in the immediate vicinity of emitting sources, there is now a widespread awareness that effects may be felt hundreds, if not thousands of miles away. The recently perceived problems of ozone depletion at the North and South Poles as a result of chlorofluorocarbon (CFC) emissions, and temperature changes attributable to emissions of carbon dioxide and other greenhouse gases, are the first examples of truly global environmental problems. The political challenges posed by pollution which crosses national borders are massive. Existing international agencies must increasingly be used as the vehicle for solutions which include as many countries as possible while addressing the diverse interests of participants.

However, not only is the geographical scale of environmental problems growing. The scope of the issues in terms of scientific complexity is also expanding. Ozone depletion, global climate change and acid rain are all issues which have required major scientific efforts to identify the problems, let alone formulate remedies. Acid rain, a regional issue rather than a global one, may be the forerunner of a number of other complex environmental problems which will require the co-operation of many countries.

Acid rain's complexity derives from, first, the way in which acid gases are transformed and transported in the atmosphere, and second, the extent to which acid pollution can cause damage to soils and ecosystems. The scientific uncertainty about these phenomena is very real and, although many links in the several chains of causality linking emissions to different types of environmental damage are now well-established (and certain chains are now complete), science cannot by itself send clear messages about appropriate remedial responses to environmental policy-makers. The scientific ambiguities have sown the seeds of confusion and mistrust which still bedevil relationships between environmental groups, governments and industry. It has been difficult to separate the scientific ingredients of policy debates from economic interests and political expediency.

In Europe, a solution, although many would argue an incomplete one, to the acid rain problem has been negotiated through the machinery of the European Community (EC) and the UN Economic Commission for Europe (UNECE). A major part of the debate concerned the completeness and the interpretation of available scientific evidence and the implications which it held for acid-rain policy. In this sense, the acid-rain story may hold valuable lessons for future debates.

Matching the scientific complexity of major environmental problems is the technological complexity of the potential solutions. Cleaning up acid emissions, for example, requires switching fuels or the installation of expensive equipment at new and existing power stations and other large industrial combustion plant. Specific regulatory solutions may create significant market opportunities for individual fuel or equipment suppliers.[1]

In this chapter, the scientific dimensions of the acid-rain problem and the technological options for reducing emissions are examined and their potential impact on regulatory solutions is assessed. It is shown that patterns of techno-logical change can influence the regulations adopted and, indirectly, the environmental problems which become salient within a particular political system.

A key question which will be raised is whether environmental policies are primarily science or technology driven. In practice, this depends on the particular issue and on the structure and interests of the institutions involved. In general, when an anticipatory approach towards environmental problems is adopted, as in the FR Germany, it appears that technological capabilities are the main driving force. However, if accepted decision-making practices coupled with economic interests create the need for a highly developed information base, as in the UK, scientific effort is likely to play the dominant role. It may be that high levels of technological capability permit anticipatory policies to emerge, while low levels of capability lead to the postponement of remedial action until problems have become intolerable.

In the following sections, the basic science of acid rain is sketched, covering both the transport and transformation of acid gases, and the damage caused to eco-systems. The more important scientific controversies are then identified. Subsequently, the sources of acid emissions are reviewed and the technological options for reducing emissions are described. The concluding sections of the chapter examine the links between science, technology and the policy-making process and the deployment of scientific and technological evidence in support of political positions.

The acid-rain problem[2]

What is acid rain?

In the strictest sense, acid rain means precisely what the term implies – rain which is more acidic than normal. The principal man-made substances which lead to acid rain in the narrowest sense are sulphur dioxide (SO_2) and nitrogen oxides (NO_x). In practice, and particularly as acid rain has become a political as well as a scientific problem, the term has taken on a far wider meaning. First, the term has expanded to cover the 'wet deposition' of all acids, whether rain has been the vehicle for deposition or not. The wet deposition of acids in sensitive areas may occur predominantly by snow or by 'occult' deposition (fog or mist).

The second sense in which the term has expanded is to leave behind any concept of rain or wetness altogether. 'Acid rain' can now cover the direct ('dry') deposition of SO_2 and NO_x as well as the deposition or effects of substances which are wholly or partly derived from them through chemical transformations. 'Acid rain' effectively covers all air pollution which derives from potentially acid-forming substances. The fact that the effects of acid rain result from many different pollutant pathways has been a source of great confusion, some accidental, some deliberate.

Acidity is measured on the pH (paper hydrion scale) which runs from zero (for a very acid substance) to 14 (for a very alkaline substance). A substance with a pH value of 7 is chemically neutral. The pH scale is logarithmic in that for every unit change of pH, the concentration of hydrogen ions present rises by a factor of 10. 'Pure' rain is actually a mild carbonic acid with a pH of 5.6 due to the chemical equilibrium between carbon dioxide in the atmosphere and that dissolved in water vapour. In practice, other substances found in the atmosphere, notably naturally-occurring SO_2, bring the pH of normal rain down further, to approximately 5, depending on the time and place.

Human activities have significantly altered the chemistry of rain. In the sparsely populated Southern Hemisphere, natural sources, such as volcanic eruptions and sea spray, account for some 90 per cent of SO_2 emissions. However, in the Northern Hemisphere, industrial activity accounts for some 70 per cent of emissions, this proportion rising to 90 per cent in Europe. Over the industrial heartlands, the pH of rain is now typically in the range 4.0–4.5, while even on the European fringes, at some distance from emission sources, a pH of 4.5–5.0 would still be typical.

Acids in the atmosphere

Table 3.1 shows the origin of the man-made SO_2 and NO_x emitted in Britain and West Germany. Most SO_2 and about half of the NO_x is emitted at some height above the ground from industrial or power-station chimneys. Most of the remainder comes from ground-level vehicle exhausts. Chimney plumes spread out down-wind of large emission sources, such as power stations, beginning to touch the ground some 5–25 km away, depending on chimney height and weather conditions. SO_2 may remain in the atmosphere for between one and five days after emission, travelling anywhere between hundreds and thousands of kilometres. The residence time for NO_x is somewhat less. Dry deposition of SO_2 and NO_x occurs directly from the dispersed chimney plume, the rate of dry deposition depending very much on the concentration of the substance, weather conditions and on the type of surface over which it is

Table 3.1 Sources of sulpur dioxide and nitrogen oxides in the UK and the FR Germany (m tonnes/year)

	FR Germany (1986)	UK (1988)
Sulphur dioxide		
Power stations	1.35	2.62
Industry	0.53	0.71
Household	0.14	0.15
Service sector	0.10	0.11
Transport	0.11	0.07
Total	2.23	3.66
Nitrogen oxide		
Power stations	0.73	0.79
Industry	0.29	0.33
Household	0.09	0.07
Service sector	0.05	0.06
Transport	1.80	1.22
Total	2.96	2.47

Source: Bundesministerium fur Umwelt, UK Department of the Environment.

passing. Uptake of SO_2 from the atmosphere is fastest over water and wet vegetation.

Acid gases are usually emitted into the atmospheric boundary layer, a layer of turbulent air close to the earth's surface within which gases are thoroughly mixed. However, at night, with a clear sky and light wind conditions, the mixing layer may become very shallow, allowing emissions from high chimneys to pass directly into non-turbulent air, where chimney plumes may be carried at very high speeds for hundreds of miles with very little dispersion.

The complex ways in which acid gases may be transported and transformed in the atmosphere can be illustrated by examining the fate of SO_2.

In the atmosphere, SO_2 is oxidized to the aerosol sulphate ion, HSO_3, at a rate of between 0.5 per cent and 5 per cent per hour. This process takes place through a reaction with the hydroxyl ion, OH. The hydroxyl ion is itself the result of a complex set of chemical reactions which can involve the creation of ozone (O_3) from the action of sunlight on man-made nitrogen dioxide, and the reaction of ozone with water vapour. The role of secondary pollutants, such as ozone, in the transformation of SO_2 to sulphate has given rise to some controversy over the extent to which a reduction in SO_2 emissions will lead to reductions in the deposition of sulphate. The so-called 'proportionality' issue and its policy significance are discussed below.

The wet deposition of sulphate from the atmosphere may occur through distinct types of process – 'rainout' and 'washout'. Rainout is believed to be the more significant source of wet deposited sulphate at some distance from emission sources, while washout may predominate at shorter ranges. Rainout involves the deposition of water droplets already containing sulphur compounds.

Washout of sulphur compounds occurs when falling rain passes through polluted air. Heavy rain, with larger droplets, is less likely to capture SO_2 or sulphate aerosol from the atmosphere in this way.

Wet deposition of sulphur compounds tends to be episodic. Measurements made at British sites have shown that about half of the annual sulphate may be deposited in only 8–9 days. This may have important implications in terms of damage effects.

Sulphur compounds are associated with about 70 per cent of the total acidity deposited at sites remote from emission sources, the remainder being attributable to nitrogen compounds. The formation of nitric acid HNO_3 from nitrogen dioxide and the hydroxyl ion is reasonably well understood. Nitric oxide (NO) is oxidized to nitrogen dioxide through a reaction with ozone, but may subsequently revert to nitric oxide through the effects of sunlight. The balance of nitric oxide, nitrogen dioxide and ozone in the atmosphere is determined by the availability of primary pollutants, by the temperature and by the intensity of sunlight. Ozone formation is therefore promoted at high altitudes and during warmer weather. This fact in itself has important implications for certain types of tree damage attributed to acid rain.

While most of the discussion has focused on the acidity of rain itself, snow and occult deposition of acidity via fog and mist may be very important at remote locations. Given that ozone formation is enhanced at high altitudes, the potential for the formation of complex pollution 'soups' involving secondary pollutants such as ozone is large and has implications for certain types of vegetation damage.

Acid-rain damage

Many types of environmental damage have been attributed to acid rain in its broadest sense. The two classes of damage which have provided the greatest spur to political action have been to lakes and streams, primarily in Scandinavia, and to forests, most prominently in West Germany. Other types of damage attributed to acid rain affect crops and vegetation, human health and building materials. While lakes and forests have been the focus of political attention during the late 1970s and 1980s, the other types of damage have been recognized for some time.

Lakes and streams

There is clear evidence that lakes and streams in Scandinavia have grown more acidic since the industrial revolution, a process which has accelerated since the 1940s. Acidification of surface waters has also been reported from mountainous areas in both West Germany and Scotland, although the low economic value of these waters has meant that little concern has been aroused. While scepticism has been expressed over the comparability of pH measurements over several decades, detailed analysis of the skeletons of diatom species (acid-sensitive creatures) deposited in the sediment at the bottom of lakes has provided conclusive evidence of long-term acidification.

This acidification has been associated with the loss of fish stocks. Research into 'surface water acidification' has therefore been aimed at three main questions – how does acidic rainfall alter the chemistry of lakes and streams, how do fish populations respond to changed water chemistry and to what extent will reductions in acid input reverse the acidification which has occurred?

Chemistry of soils and surface waters

The bulk of precipitation entering a lake (80–90 per cent) passes first through the soil in the catchment area, where chemical reactions alter the composition of run-off water. Even if the rain entering the soil were pristine, chemical processes within the soil, some natural, some influenced by soil cultivation, could result in acidic run-off.

Changes to the acidity of soils are the result of acid-producing and acid-consuming processes. The uptake of base cations – essential plant nutrients – from the soil is an important acid-producing process. In completely uncultivated conditions this would not matter as when plants died the cations would be returned to the soil, maintaining a long-term equilibrium. However, if plants are harvested, the acidification added to the soil becomes permanent.

Another acid producing process is the leaching of organic acids from plant litter. This process is particularly important for coniferous trees which have been identified as a major cause of soil acidification.

The weathering of bedrock material underlying soil systems is a major acid-consuming process. Bedrock is available in enormous quantities in comparison to annual deposited acidity, a 1 mm layer being able to absorb hundreds, or even thousands, of years of hydrogen ion deposition at current rates. However, bedrock weathering is limited by the rate of the chemical reaction and the limited amount of soil moisture which ever comes into contact with it.

Sulphate adsorbed in soil, by either chemical or biological processes, is also an acid-consuming process. If sulphuric acid deposition declines, the adsorption process can be reversed and sulphate and hydrogen ions may be released back into the soil, partially compensating for reduced acid deposition. The time it might take for acidification to be reversed is determined by the stock of soluble sulphate in the soil.

Acidification processes are often of an episodic nature, leading to acid 'flushes' which can have severe implications for fish life. These episodes are generally associated with heavy rain or a sudden snow-melt, which follows a long dry period. The spring snowmelt leads directly to very acid runoff as most of the acidity deposited over the winter season is released very quickly as melting begins.

A further important aspect of the interaction between deposited acidity and soils is that hydrogen ions may be exchanged for aluminium ions which can be toxic to both plant and fish life.

Acid rain and fish life

Early concern about acid rain in Scandinavia was driven by fairly anecdotal evidence of a link between fish loss, the acidity of lakes and the acidity of rain. Subsequent scientific investigation, while broadly confirming the hypothesis, has revealed a more complex chain of causality. Not only acidity, but also the concentrations of calcium ions and aluminium-based compounds are significant determinants of fish survival. Calcium and aluminium ions are leached from surrounding soils as a result of deposited acidity.

The effects of the three causative factors – pH, calcium and aluminium – are not simple and cannot be considered in isolation. Nevertheless, higher aluminium concentrations and lower calcium concentrations appear to be strongly linked to fish death. There is a more moderate direct link between low pH and fish deaths, although waters with pH below 4.0 appear incapable of supporting fish life.

As pH falls, the concentration of both aluminium and calcium ions increases. The increase in aluminium ion concentration will prejudice fish survival, while that of calcium ions will enhance it. This suggests that a complete strategy for rehabilitating acidified lakes would involve a reduction of acid deposition coupled with additions of lime to the catchment area to boost calcium ion concentrations. Liming alone is not a sufficient remedy.

Water with altered chemistry has different effects on fish at different point in the life cycle. Acidity may reduce the ability of fish to spawn, inhibit the ability of fertilized eggs to survive as well as killing adult fish. Fish eggs may be particularly vulnerable to acidified water. This is reinforced by the tendency for acid 'flashes' during the spawning season caused by the spring snowmelt. Acid episodes are believed to be a particular hazard to fish life.

Forest damage

Very high concentrations of SO_2, such as those found near major industrial sources, have long been known to cause direct damage to trees, involving needle browning and loss and ultimately tree death. Tree damage of this type is still widespread in Czechoslovakia and East Germany and was observed in the vicinity of the Sudbury smelter in Ontario, the largest source of SO_2 in the Western world.

In the early 1980s, a new type of tree damage was observed in different parts of West Germany. The wide range of symptoms included reduced tree growth, increased mortality, yellowing and premature loss of needles, the thinning of canopies and changes to fine root systems. Damage to fir, spruce and pine trees was initially detected but has also been observed on hardwoods, notably beech, birch and oak trees.[3] The symptoms have been observed in a very wide of situations including natural and planted forests, acid and basic soils and soils of varying degrees of fertility. While older trees have been worst affected, damage has been seen in trees of all ages. Damage has tended to be worst on west or north-west facing slopes at higher altitudes. It has now been established that there are several types of damage syndrome, all of which have become apparent at roughly the same time.

Similar damage effects have subsequently been observed in France, the UK, Southern Sweden, the Netherlands, Italy, Austria and Switzerland. Given the cultural and economic value of German forests, forest dieback (Waldsterben)[4] has been most intensively researched and documented in West Germany. The health of German trees has been documented in a series of forest damage surveys conducted by official West German agencies since 1982, using foliage loss as an indicator of tree damage. The proportion of German forest area in the moderately and severely damaged categories (more than 25 per cent foliage loss) was 15 per cent in 1983, stabilizing at about 20 per cent in 1985 and 1986. The earlier surveys included trees with more than 10 per cent foliage loss in the damaged category, which produced estimates of damaged forest areas of 34 per cent in 1983 and 56 per cent in 1986. Trees with 10–25 per cent foliage loss are now referred to as being at the 'warning stage'. These earlier high estimates of forest damage did much to fuel public anxiety about Waldsterben.

The scientific understanding of forest damage is much less complete than that of surface water acidification. A variety of hypotheses, competing and complementary, exist to explain the various types of forest damage which have been observed. An early hypothesis advanced by a German soil scientist, Bernard Ulrich, held that soil acidification was the underlying cause of poor forest health, as aluminium leached from the soil damaged the fine root system of trees and impeded the uptake of vital nutrients such as calcium, magnesium and potassium. The Ulrich hypothesis gained widespread currency in West Germany during the early 1980s.

The corollary of the Ulrich hypothesis is that deposited acidity, and hence SO_2 emissions, were largely to blame for forest damage. This led to power stations and other large emission sources being identified as the main culprits. Subsequently, however, it became apparent that the Ulrich hypothesis could not explain the observed patterns of forest damage which tended not to be correlated with soil acidity, being most pronounced at higher altitudes on exposed ridges.

Many other theories have been advanced. These include: interactions between ozone and acid mist acting directly on trees; magnesium deficiencies in soils caused by tree harvesting and leaching as a result of acid deposition; excess nitrogen uptake of nitrogen caused by the escape of ammonia from soil fertilization (noted in particular in Belgium, the Netherlands and Northern Germany); and the 'multiple stress' hypothesis.[5]

The latter hypothesis is that air pollution (ozone, acid mists and acid precipitation) may trigger severe damage in trees which have already been stressed by soil acidification or by extremes of weather, such as drought or severe frost. Multiple stress does not constitute a single hypothesis, but establishes a flexible scientific framework for examining the several types of damage syndrome which have been identified. Some scientists are sceptical of the value of the multiple stress framework because its very broadness makes it untestable.[6] After falling out of favour because of greater emphasis on the role of ozone, the Ulrich hypothesis is seen as a component of the multiple stress concept.

Factors playing a role in the terminal phase of tree damage include fungal infections and insect infestation. However, these are recognized as providing the coup de grace for trees already subjected to other stresses. Tentative damage hypotheses which have been dismissed include forest management, long-term climatic change, radioactivity effects and viral infections.

While forest damage is far from being completely understood, a consensus is emerging that air pollution may be but one factor among many contributing to a complex set of damage syndromes. Air pollution may be a predisposing, inciting or contributory stress. Considerable progress has been made in developing computer models which simulate the effects of pollution on tree health. These models, which may apply to individual trees, to tree stands or to whole ecosystems, are helping to improve understanding of the mechanisms involved when multiple factors play a role in tree damage.

The Forestry Commission in the UK has commissioned its own tree damage surveys,[7] two years behind Germany, in response to strong pressure from Friends of the Earth which had commissioned its own independent survey in 1985. Comparisons show that, for some species, British trees, notably Norway spruce and deciduous trees, such as beech and oak, are as damaged as those in Germany. However, the British public has not responded with the same degree of anxiety as Germans, partly because the German tree damage estimates were initially exaggerated, and partly because of the lower cultural and economic importance of forests in the UK.

Other acid-rain damage

The most plausible type of effect on crops and other forms of vegetation is through changes to growth patterns. Few acute effects have been observed. There is evidence from fumigation experiments that high ambient concentrations of SO_2 may inhibit the growth of certain crops. However, many experiments have been conducted with SO_2 concentrations rarely found in practice. More recently, there has been some concern about the effects of ozone on crop growth, particularly in combination with SO_2. Lichen populations are highly sensitive to SO_2 concentrations, and the particular species present at a location are a reliable indicator of ambient air quality.

The direct effect of high concentrations of SO_2 on human health has been well-known for decades. The 4000 deaths through chest disease in London following the smog of 1952 are attributed to smoke and SO_2 acting together. In Europe, North America and Japan, significant reductions in urban concentrations of particulate matter and SO_2 have been obtained by switching to cleaner fuels, fitting grit arrestment to solid-fuel fired plant and building higher chimneys to disperse SO_2 emissions. Some controversy remained over the role of low-level concentrations of sulphate in damage to human health. The accepted view in Britain (and that embodied in World Health Organisation guidelines) is that damage to health does not occur below a threshold of about 50 ug/cum of SO_2. However, in the United States and Japan, concern about the possible health effects of low-level concentrations of SO_2 was an important factor in determining requirements to fit flue gas desulphurization equipment to new power stations during the 1970s.

More recently, there has been concern in Scandinavia about the health effects of toxic metals leached from the soil and pipes by acid water. There is little concern about acid drinking water in general, but deaths through high aluminium concentrations encountered during kidney dialysis have been suggested in Sweden.

It is well-known that SO_2 deposited on certain types of stone, notably sandstone, limestone and marble, cause considerable damage through chemical reaction. In addition, the higher volume of the calcium sulphate formed may cause physical damage through surface flaking. Building damage in urban environments is primarily the result of local sources of pollution which, in Britain and many other countries, have fallen significantly in recent decades. Continuing damage to stone may be the result of SO_2 deposited some time ago in the cracks and interstices of stone. Nitrogen compounds acting together with bacteria may also damage materials.

Scientific controversies over acid rain

The acid-rain debate at the scientific level has been characterized by a number of disputes, many of which have subsequently been settled. This scientific uncertainty allowed both advocates and opponents of emission abatement measures to justify their positions with respect to science. The tendency for the interpretation of partial scientific evidence to coincide with the interests of individuals and organizations will be discussed at the end of this chapter.

The controversies have covered the transport and transformation of pollutants (the 'how far does it travel' and the 'proportionality' issues), the reversibility of damage and the causes of fish mortality and forest damage.

How far does SO_2 travel?

This issue was raised, primarily by electric utilities and other emitters, in the 1970s. The general objection was that 'there was no direct evidence' that SO_2 in the atmosphere travelled far enough to pollute Scandinavia say, after having been emitted in the UK. Objections of this type have been laid to rest by airborne measurements of tracer gases in power station plumes and by the pollutant transport modelling carried out under the UNECE EMEP Programme. However, the electricity supply industry in the UK has derived some satisfaction from the fact that recent revisions to EMEP results have reduced the estimates of the British contribution to acid deposition in Norway and Southern Sweden.

Proportionality

The question of whether emissions reductions will lead to proportional reductions in deposition is of more substance. While, at the global level, what goes up must come down, the relative proportions of SO_2 deposited in the wet and dry forms, and close to or remote from the emission source may vary according to the quantity of SO_2 emitted. This is because, as described above, other substances, notably ozone, play a role in the chemical conversion of gaseous SO_2 into aerosol sulphate which may be deposited as 'acid rain' in the narrowest sense. As the rate at which SO_2 is converted to sulphate may be limited by the availability of atmospheric ozone, it is possible that a reduction in SO_2 emissions may not lead to a proportional reduction in wet deposition of sulphuric acid. However, it has not been suggested that SO_2 emissions reductions will lead to a zero reduction in wet deposition.

The UK Meteorological Office and the CEGB have made considerable progress in resolving the proportionality issue. Using relatively simple atmospheric models, they have shown that from diffuse SO_2 sources (comprising many low-level emitters in an urban location), wet deposition is proportional to emissions at distance greater than about 200 km from the source. From single point sources (such as an isolated power station), the comparable distance is 500 km, while, for groups of large sources (such as the clusters of power stations found in Yorkshire and the Midlands in Britain, and the Ruhr in Germany) non-proportionality may extend up to 1000 km, tantalizingly close to the distance between Southern Scandinavia and the industrial heartlands of Britain and Continental Europe. Nevertheless, this work shows that significant, if not precisely proportional, deposition reductions in sensitive regions of Scandinavia might be expected from SO_2 reductions.

Damage effects

The understanding of damage effects is far from complete, leaving considerable scope for differences of scientific interpretation. The role of sulphate and calcium 'stocks' in the soils of the catchment areas of acidified lakes is an example of this.

The CEGB in Britain had placed considerable emphasis on the fact that soils can contain quantities of sulphur, locked up in various chemical forms, which are equivalent to tens, if not hundreds, of years of wet deposited sulphate.[8] Sulphate, accompanied by hydrogen ions, can leach from the soil if acid deposition is reduced, so that, at least initially, lower acid deposition does not necessarily lead to lower acid runoff. The CEGB has inferred from this that it may take many decades before emissions reductions are reflected in improved surface water quality. Some Scandinavian scientists, on the other hand, disagreed and pointed out that much of the sulphur in soil is locked up in non water-soluble form and cannot be leached out in responses to lower soil acidity.[9] This implies that the recovery time for acidified waters may be much more rapid than the CEGB has suggested. The results of experiments under way, which may last many years, should provide a reasonably definitive answer to this question.

The different approaches adopted for the interpretation of incomplete scientific information, and the implications for responding to the acid-rain problem at the political level are discussed below.

A much greater degree of certainty has recently emerged about the levels by which acid deposition would need to be reduced in order to protect ecosystems.

The UK Acid Waters Review Group[10] has concluded that: 'to maintain current conditions, a deposition reduction of around 30% from 1985 levels is necessary'; and that 'a reduction of 90% from 1985 deposition levels would be required to return most surface waters to near-pristine conditions'.

Clean-up technology

As shown in Table 3.1, power stations and industry are responsible for a very large proportion of the SO_2 emitted in Britain and Germany, and for a significant proportion of the NO_x, although transport is the largest source for the latter pollutant. The combustion of coal is largely responsible for SO_2 and NO_x emissions from stationary sources in both countries. A variety of methods exist for burning coal more cleanly, either through end-of-pipe control technologies, such as flue gas desulphurization (FGD), selective catalytic reduction (SCR) of NO_x, or through processes which are intrinsically less polluting.

Fuel characteristics

SO_2 from the combustion of fuels derives entirely from sulphur contained in the raw fuel. Worldwide, the sulphur contents of coal may vary from 0.5 per cent to 5 per cent. If the calorific value of the coal is very low because of a high moisture or ash content, the quantity of sulphur emitted per unit of heat produced may be very high. This may be particularly true of brown coal (lignite) which, chemically, is mid-way between peat and hard (bituminous) coal. The sulphur content of coal burned in British power stations varies between 0.5 per cent and 2.0 per cent with an average of 1.6 per cent. German hard coal has a slightly lower sulphur content. Around 5 per cent of the sulphur in coal is retained in ash rather than being converted to SO_2. The disposal of ash and FGD wastes can lead to other environmental problems.

The sulphur content of fuel oil which is burned in power stations or industrial boilers can vary from 0.2 per cent up to 4 per cent depending on petroleum refinery operations and the sulphur content of the crude oil feedstock. In Britain, prior to the production of North Sea oil, the typical sulphur content of fuel oil was around 3 per cent, which meant that fuel oil gave rise to more SO_2 per unit of heat than did coal. Sweeter North Sea crudes, combined with the greater use of cracking facilities to convert fuel oil into higher value products such as petrol, has resulted in the average sulphur content of fuel oil falling to about 2 per cent during the 1980s. As a result, the combustion of fuel oil now results in lower SO_2 emissions than does coal. In many parts of Europe, the typical sulphur content of fuel oil is now only 1 per cent, partly as a result of environmental regulation.

Natural gas is essentially sulphur free.

Nitrogen oxides (NO_x) emitted during the combustion of fossil fuels arise both from the intrinsic nitrogen content of the fuel and from the oxidation of nitrogen in combustion air. NO_x emission levels are therefore a function of the way in which fuel is burned and can be rather variable. For a large power station boiler, NO_x concentrations in flue gases might be of the order of 600–900 parts per million (ppm) for coal, 350–500 ppm for oil and 250–350 ppm for gas.[11]

Fuel cleaning

It is possible to treat both coal and fuel oil so as to reduce the sulphur content. As much as half of the sulphur content of coal is in pyritic form which can be removed by relatively simply physical processes, exploiting the different densities of combustible material and ash. Coal cleaning is commonly used to reduce the ash content of coal. Such sulphur removal can be obtained relatively cheaply, for a few £ per tonne of coal, in association with other benefits such as reduced ash contents. Removing the remaining organic sulphur from coal would require chemical or biological processing at a much higher cost.

The desulphurization of fuel oil in refineries is also possible, but at a considerably higher cost. In practice, the sulphur content of fuel oil has declined in recent years in association with the higher proportion of crude being converted into higher grade products such as petrol, aviation fuel and gas oil. To achieve this 'lightening of the barrel', fuel oil has been converted in cracking plant where sulphur is captured and may be recycled for subsequent use. This trend has led to significantly reduced emissions from fuel use in refineries themselves as well as from other fuel oil users.

Combustion modification

Modification of combustion processes can secure significant reductions of NO_x emissions from large boiler plant – up to 70 per cent reductions have been secured, although reductions of 40–50 per cent would be more typical. Where combustion modification techniques are retrofitted to existing boilers, smaller emissions reduction of the order of 30–40 per cent might be expected. The essential feature of all methods of combustion modification is that they reduce combustion temperatures, and reduce the time during which fuel is exposed to elevated temperatures.

A variety of specific techniques of varying sophistication may be used. These vary from lowering the amount of excess combustion air used, through to recirculating flue gases, carrying out combustion in two stages and the use of true 'low-NO_x' burners. Combustion modification is relatively cheap, costing only a few £/kW of installed electricity generating capacity compared with a total investment of up to £700–800/kW. Low-NO_x burners are accepted as the best available technology for NO_x reduction in the US and in most of Europe. However, in Japan and in the FR Germany, more expensive flue gas cleaning processes are required.

Flue gas cleaning

Flue gas cleaning is the only real option for reducing SO_2 emissions from fossil fuel combustion, and the only option which will reduce the greater proportion of NO_x.

SO_2 removal

Modern and effective FGD designs plant were developed in Japan and the US during the 1970s. The FGD principle had been long established, but early designs, used in the UK for example (as described in Chapter 9), were crude and unsatisfactory in terms of reliability, performance and secondary environmental impacts.

150,000 MW(e) of FGD has now been installed worldwide, of which 90 per cent is situated in three countries – the US, Japan and the FR Germany.[12] Most FGD plants remove more than 90 per cent of the SO_2 contained in flue gases. While there are many types of FGD process, the most common process is the wet lime/limestone scrubber which accounts for 75 per cent of total capacity installed. A wet slurry of lime or limestone is sprayed through the flue gases, reacting with the SO_2 to form a calcium sulphite sludge plus carbon dioxide. The calcium sulphite sludge may be disposed of directly, as is predominantly the case in the US. However, because of the secondary environmental impacts of sludge disposal and shortage of land, the calcium sulphite is generally oxidized to form gypsum (calcium sulphate) in Japan and Europe. Gypsum may then be sold for use as an aggregate or for use in manufacturing wallboard. For every 100 tonnes of coal burned in a power station fitted with FGD, approximately 6 tonnes of limestone are required and 10 tonnes of gypsum are produced. Environmental problems associated with the mining of limestone and the disposal of solid and liquid wastes from FGD have become local environmental issues in Britain and Germany.

One of the main alternatives to the wet lime/limestone processes is the Wellman Lord regenerable FGD system which recovers elemental sulphur or sulphuric acid by using a sodium sulphite reagent. As the reagent is regenerated, the raw material requirements are much lower for the Wellman Lord system. However, the investment costs are higher which makes it less attractive to power station operators.

In general, FGD will add ~15 per cent to the basic cost of a modern conventional coal-fired power station. Costs will be higher if FGD is to be retrofitted to an existing power station. The German 37 000 MW(e) FGD retrofit programme cost 380 DM/kW, while estimated British retrofit costs are approximately £150/kW.

Operating costs for FGD are also considerable. The cost of electricity produced from a power station fitted with FGD will be approximately 15–20 per cent higher than at an uncontrolled station.

As well as the direct costs, electricity costs are incurred because the efficiency of power stations is lowered. In Japan and Europe, 2–3 per cent of the output of the power station is needed to drive FGD fans and pumps. The higher figure would generally be reqired where FGD has been retrofitted because, in this circumstance, flue gas ducts tend to be longer. In the US, steam from the boiler is used to reheat flue gases which have cooled during the scrubbing process and as much as 5 per cent of station output may be lost. In Europe, flue gases are reheated using heat exchangers.

NO_x reduction

NO_x reductions of up to 80–85 per cent can be achieved by using selective catalytic reduction (SCR) processes. These use a combination of ammonia and catalytic honeycombs or plates to reduce NO_x. SCR is simple, but expensive, as the catalysts need to be replaced every 4–5 years. The installation cost for SCR in Germany was 210 DM/kW spread over 33 000 MW(e) of capacity. It is only in Germany and Japan that SCR is regarded as best available technology because of its high cost and the availability of relatively effective combustion modification. Even in the FR Germany, SCR is required only at hard coal-fired power stations, not at those fired on lignite, oil or gas.

Clean generation technologies

At present, one of the most attractive options for power generation is combined cycle gas turbine (CCGT) plant. Given the present low price of natural gas, and the lower capital cost of CCGT plant, it is unlikely that conventional coal-fired power generation will take a major share of power plant orders over the next few years. Such stations, with an overall conversion efficiency of, at best, 40 per cent and with the need to fit expensive flue gas cleaning equipment may look even less attractive in the context of the carbon dioxide contribution to global warming.

However, after the 1973 oil crisis, considerable R&D efforts were put into developing cleaner, more efficient ways of burning coal. These are now beginning to bear fruit. As a result a cluster of new technologies, falling into the generic category 'clean-coal technology', is developing. These may be available on the market in the late 1990s.

A general feature of all the new technologies is that they rely on both a gas turbine and a steam turbine for generating power. Consequently, conversion efficiencies are expected to be much higher than in conventional coal-fired power generation. In addition, the new technologies are modular and may be installed in smaller capacity increments. The escalating construction times and interest payments associated with large, conventional coal-fired power stations have begun to make them unattractive to utilities faced with electricity demand growing at a relatively slow rate.

There are a number of different variants of fluidized bed combustion (FBC) technology. In these, coal is burned in a bed of fluidized material, either at atmospheric or elevated pressures. As a result of lower combustion temperatures, NO_x formation is intrinsically lower than in pulverized fuel firing. Also, limestone or dolomite may be added directly to the fluidized bed in order to secure sulphur removal in situ, removing the need for flue gas cleaning processes. Worldwide, some 430 fluidized bed boilers are now operating.[13]

Coal gasification allows higher efficiencies and lower emission characteristics. The gasification process leads to the formation of hydrogen sulphide which is much easier to remove than SO_2. The 'topping cycle' process promises considerably higher efficiencies, approaching those of CCGT plant, but is unlikely to be commercially available until the late 1990s.

The result of these technological developments is that competition for power generation systems in the future is likely to be between CCGT and clean coal technologies. Effectively, FGD will be used as a retrofit technology which is used to make an older design of power stations more environmentally acceptable. New nuclear power stations are unlikely to be constructed in either Britain or Germany for some time to come given high costs, partly caused by licensing hurdles.

The role of expert advice

The science/politics interface

Scientists and politicians have very different roles to play.[14] They see themselves and the ways in which they influence the world in very different ways. Scientists tend to place a high value on objectivity, the search for truth and, when it comes to decision-making, the quest for the 'right' or 'optimal' solution. Politicians, on the other hand, live with the axiom that politics is the art of the possible and are much more familiar with ambiguity, compromise and settling for the second best. Politicians may need 'objective' facts to legitimate expedient decisions.[15] There is thus a very real tribal suspicion between politicians and scientists.

This book focuses not on ideal solutions, but on the social and institutional factors which shaped the actual responses to acid rain in Britain and West Germany. The gap between retrospective analyses of political decisions and presciptions of optimal policy solutions arrived at through economic and scientific analysis is wide. Clearly, from the latter point of view, politicians and administrators can appear 'irrational', if not perverse, in the policies which they choose and implement.

Yet, the very concept of irrationality begs many questions. The rationality of the 'optimal' solutions predicted from economic theory is often based on far too narrow a characterization of the decision-making space. In the context of modern pluralistic democracies, those in policy-making circles are dancing to many tunes in addition to those played by scientists and economic analysts. Environmental policy is not established in a vacuum but is influenced by and, in its turn, has implications for energy policy, industrial policy and even foreign policy. Depending on the issue, public opinion, or the opinions of key opinion formers (Members of Parliament, senior civil servants), may wield a strong influence.

The party (or parties) in power in a country may have a firmly established political agenda on which environmental issues may or may not occupy a prominent place. If environmental issues have a low priority, the implementation of a particular environmental protection measure will depend crucially on the extent to which it can be integrated and is compatible with the government's principal policy objectives.

The analysis of the acid-rain debates in Britain and West Germany suggests a suitable metaphor for the role of science and economics. Prescriptive, rational arguments have effectively played the role of the intellectual artillery in the political conflicts between groups with different interests in the outcome of the acid-rain debate. In spite of the very different outcomes in Britain and Germany, the artillery used, i.e. the scientific arguments advanced on each side, were of a similar nature. What differed between the two countries was not the quality of the intellectual arguments but the relative political strengths of the groups advancing them.

Although the importance of scientific analysis in determining the outcome of the political debate over acid rain has not been dominant, it is necessary to give further consideration to the use of scientific argument. This is partly because the rhetoric of proper scientific understanding attained such a prominent place in Britain's international negotiating position and because, in the context of the less pluralistic decision-making structures found in British government, greater scope was provided for scientific arguments to come to the fore.[16]

Also, in Britain, the greater concentration of political power, exercised in the context of mild public interest in the acid-rain issue, allowed official policy to be determined by a comparatively small, elite group of people. Most of these, unusually for Britain, happened to have professional scientific training. Thus, while the interpretation of scientific evidence was one component of the acid-rain debate in West Germany, it became a more critical element in Britain.

The communication of scientific evidence

Scientists are often portrayed as dispassionate seekers after the truth. At the level of the individual experimenter, this may well be the case. However, the operation of science as a large-scale enterprise, through all the stages of determining research priorities, deciding budgets, allocating funds to specific institutions and, finally, actually conducting the scientific work, has the same dimensions as any other business activity. The choice of which lines of inquiry to pursue and the identity of the groups which actually carry out specific projects may affect the way in which the scientific findings are interpreted and promulgated outside the scientific community.

A further problem is the process by which the state of scientific knowledge is transmitted to politicians, administrators and the interested public in a politically controversial area such as acid rain.[17] Scientific literature is highly specialized and often impenetrable to the outsider, even to scientists trained in other disciplines. Thus, for an issue which requires the participation of a wide range of disciplines, from atmospheric chemistry and physics through soil science to biology, a necessary first step is the production of scientific review papers, which summarize the key results of experiments from several different disciplines, identifying the current state of knowledge and the most important unresolved questions.

Many reviews of acid rain have been produced by governments, by industry and by interest groups. Even then, politicians, top businessmen and the public do not gain their understanding from scientific review papers. A further layer of sifting inevitably takes place. Most top decision-makers must base their judgements on information which has been condensed into no more than a few sheets of paper. Equally, the interested layman will generally base his judgment on, at the most, a few 1000-word articles in newspapers or journals.

This multi-layered process of condensing and sifting information, because it is essentially interpretative, is virtually impossible to do in an entirely neutral way. Every decision about what to include and what to omit is judgemental in nature. It is very clear that different people have produced review papers on acid rain which imply very different political responses. It is indeed quite clear that interpretations tend to reinforce the interests of the organizations sponsoring the reviews. This is plainly borne out by comparing the surveys carried out by groups as diverse as the CEGB, Friends of the Earth and the Nature Conservancy Council in the UK or the Green Party, the Umwelt-bundesamt (Federal Environment Office) or VDEW (Association of German Electricity Producers) in Germany.

The risks of science being compromised have been recognized both by the Department of the Environment (DoE) and the electricity supply industry in the UK. The DoE has sponsored a number of independent acid-rain review groups which have been free to reach their own, often forthright, conclusions. Research sponsored by the former Central Electricity Generating Board was run through a blind trust administered by the Royal Society and its Scandinavian counterparts.[18]

The phenomenon of sifting and selectivity is reinforced by the multi-layered process of review and interpretation. At each stage, the information retained is likely to appear more conclusive, with countervailing indications and results being suppressed.

That is not to say that the individuals involved in the review process are making cynical use of the available material. When vast quantities of scientific literature are being scanned, it is essential to identify and abstract key 'facts'. The selection of these facts and the extent to which they appear pertinent to the reviewer may well be conditioned by prior attitudes and the unconscious anticipation of the degree of acceptance within an organization.[19]

Science is now a collective rather than an individual activity. The process of group dynamics is likely to reinforce certain styles of research and interpretation of results among closely knit and relatively self-contained research communities such as exist in the laboratories of governments and large industrial organizations. Given the way that scientists in these organizations must interact with politicians and administrators, it is obvious that ideal conditions exist for conventional wisdoms and house styles and attitudes to emerge. Individual scientists who find themselves at odds with the group consensus may find it very hard to argue, or even express, a dissenting view.

Strategies for the interpretation of evidence

The process of reviewing complex scientific evidence in a politically contentious field such as acid rain has led to a number of distinct strategies for interpretation which may be employed to support, albeit unconsciously, a particular viewpoint.

The first, used particularly by environmental pressure groups is simplification. This can easily be used to justify remedial action as a matter of urgency. In the context of acid rain, this approach would tend to ignore the many different pathways linking the emissions of different pollutants with the many different types of environmental damage caused. Thus, 'acid rain' is used as a universal concept linking any type of emission with any effect, regardless of scientific evidence about specific mechanisms for damage.

On the other side of the debate, industry and reluctant governments have tended to use the 'incongruity' strategy, exploiting the complexity, scope and incomplete nature of scientific evidence.[20] This involves a focus on the detail of scientific results, and an emphasis on findings that are apparently counter-intuitive. 'Incongruities' which have been emphasized include the correlations (or apparent lack of them) between acid emissions and deposition, the quantities of sulphate deposited in air masses which have not passed over pollution sources and the lack of correlation between acidity and fish death in lakes of certain chemical qualities.

Two further, related strategies used by those opposing action to curb acid rain may be conveniently labelled as the 'direct evidence' and 'burden of proof' arguments. Indications of environmental damage may be qualified by the caveat that the evidence presented is circumstantial rather than direct, implying that circumstantial evidence is inadmissible as a basis for remedial action on acid rain.

However, a problem like acid rain must be simplified by breaking it down into its component parts – emission sources, atmospheric physics, soil chemistry and biological effects. When reviewing the problem in the round, it is realistic to consider each part of the whole system as though it were independent of the others, carrying only the most basic ideas and items of information over from one link of the causal chain into the next.

This necessary process of simplification is not conducive to the turning up of direct evidence. In fact it makes it, to all intents and purposes, impossible. Thus, establishing causal links between emissions and environmental damage is not a matter of unambiguous deduction from known facts, but one of inference, a more elusive and subjective process. 'Direct evidence' for acid-rain damage is, therefore, essentially impossible to obtain.

The concept has been used as a challenge to those advocating stricter emission controls. This effectively allocates the burden of proof on to the control advocates, or, more strictly, presenting them with the impossible burden of clarification where circumstantial evidence already indicates a connection between emissions and damage.

From the evaluation of science to rational policy?

The challenge of producing a rational response to environmental problems goes beyond the search for conclusive scientific evidence, taking us into the domain of economics and decision science. A recent report commissioned by the UK government[21] recommends the greater use of economic techniques and insights in developing and implementing environmental policies. In principle, a rational allocation of resources to environmental protection can occur only if scientific assessments are located within a secure framework of economic analysis.

One of the principal building bricks in the economic assessment of environmental policies is the technique of cost-benefit analysis. This essentially involves the assignation of monetary values to both the costs of environmental controls *and* the resultant environmental benefits. The technique can be controversial and the appropriateness of assigning monetary values to assets such as the fabric of medieval cathedrals, the life-supporting capacity of remote lakes and the health of trees which are not used for commercial purposes can be questioned.

Economists have developed a number of techniques for assessing the values of environmental assets, including *contingent value methods* and *hedonic pricing*. Contingent value methods essentially involve polling people about their willingness to pay for the availability, or even the mere existence, of environmental assets. There are very few assets which cannot, in principle, be valued by this method. However, the polling techniques usually ask people to place themselves in hypothetical situations. It may not always be the case that actual willingness to pay in a market place would be the same as the theoretical willingness to pay elicited in response to a survey.

The principle of hedonic pricing is to infer the valuation of environmental assets from actual changes in related markets. For example, the cost of aircraft noise might be deduced from changes in the value of house prices close to an airport. There are potential applications of this technique in the acid-rain field. The value attached to lakes which are well-stocked with fish might, for example, be deduced from the prices paid for fishing permits.

Cost-benefit analysis appears to have played little role in determining the course of the acid rain debate, partly because of the difficulty of pricing environmental costs and partly because of potential abuses of the technique. In the preamble to its proposals for a Large Combusion Plant Directive,[22] the Commission of the European Communities suggested that environmental damage costs of up to $3.5bn/year might be attributable to acid rain. This number was extremely uncertain, and was not a measure of the damage which would have been avoided had the Commission's legislative proposal been put into effect. The weaknesses in the Community's cost-benefit arguments were criticized in the UK,[23] and undoubtedly made agreement of control measures more difficult.

In its purest form, cost-benefit analysis takes the form of a balance sheet where costs and benefits are assigned single values. It is implicitly assumed that the consequences of environmental protection measures are known with certainty. For most environmental problems, this is not the case.

A potential way out is offered by decision analysis, a technique which has become popular in the management science community. Within this framework, the uncertainty inherent in a problem such as acid rain can be taken into account explictly by attaching probabilities to the possible consequences of different courses of action. It is not always possible to identify objective probabilities of given outcomes. Therefore, 'subjective' probabilities are frequently elicited from a group of experts in the appropriate fields.

In the more sophisticated decision analysis models, the acid-rain phenomenon might be broken down into its different components, and different subjective probabilities attached to the outcomes at each step. The way in which probabilities propagate along the causal chain can be assessed in these sophisticated models.

Decision models can incorporate all the axioms of cost-benefit analysis and can therefore be used to monetarize the potential environmental benefits of emission controls. Operated in this mode, decision models can, like straight-forward cost-benefit analyses, offer direct prescriptions for action based on expected or most likely outcomes (albeit prescriptions that are heavily value-laden).

However, decision models can also be operated so as to describe environmental benefits purely in terms of fishless lakes, dead trees or other suitable indicators, without attempting to assign monetary values. Decision models used in this way offer the opportunity to test the robustness of the expected environmental consequences of specified emission controls given some quantification of the incompleteness and uncertainty inherent in current scientific understanding.

Decision models have been applied to the acid-rain problem by a number of bodies including the Commission of the European Communities, the UK Department of the Environment and the former Central Electricity Generating Board. However, it is not evident that the use of decision models has had a significant impact on the policies of these bodies. Senior decision-makers at the political level have not been directly influenced by the results of studies. However, decision models have been used by civil servants and by support staff. It may be that the use of both cost-benefit analysis and decision models has clarified issues when advice was being prepared for senior decision-makers.

Decision-making in practice

There appear to be two fundamental constraints on the use of techniques such as cost-benefit and decision analysis in solving environmental problems which are the subject of political controversy. First, they rely on the highly controversial monetary valuation of assets. In pluralistic democracies, 'true' values may only be revealed through the operation of the political process itself, as different social groups bargain and arrive at acceptable compromises which implicitly place values on environmental goods. The use of *a priori* values derived by technocratic elites might short circuit lengthier, but legitimate, political processes.

Second, all models are abstractions from reality, and this is no less true for decision models. The characterization in any model of the choices available to individual decision-makers are inevitably simple.

In practice, the links between scientific evidence and political action are elusive, probably too subtle to be captured in the simplified frameworks necessary in creating even the most sophisticated decision models. Almost the whole of this book is devoted to describing decision-making structures. It will become clear that it is practically impossible to identify clearly who the 'decision-makers' really are. An enormous variety of groups and individuals, in and out of government, helped to determine the very different decisions made on the acid-rain question in Britain and Germany. These influences are the topic of Part II of the book.

Notes

1. H.J. Leonard (1988), *Pollution and the Struggle for the World Product*, Cambridge University Press, Cambridge.
2. The discussion of the science of acid rain has been built up from a variety of sources including: Environmental Resources Ltd (1983), *Acid Rain: A Review of the Phenomenon in Europe*, Graham & Trotman, London; Stockholm Environment Institute (1990), *Air Pollution in Europe: Environmental Effects, Control Strategies and Policy Options*, Stockholm; UK Acid Waters Review Group, December 1988, *Acidity in UK Fresh Waters*, HMSO, London; UK Terrestrial Effects Review Group, 1988, *The Effects of Acid Deposition on the Terrestrial Environment in the UK*, First Report, HMSO, London; Blank, L.W. *et al.* (1988), New perspectives on forest decline, *Nature*, Vol. 336, No. 6194, 3 November 1988, pp. 27–30; and *CEGB Research* (1987), Acid rain, a special issue, No. 20, August 1987.
3. K.F.A. Juelich (1986), *Waldschäden: Ursachenforschung in der BRD und den USA*, Bonn; and E-D. Schulze (1989), Wie kann dem Wald geholfen werden, *Forschung*, 3/1989, 17–9.
4. The dramatic term 'Waldsterben' has now been supplanted by the milder term 'Waldschäden' (forest damage) in official German reports.
5. P. Schütt and E.B. Cowling (1985), Waldsterben: a general decline of forests in Central Europe, *Plant Disease*, 69 (7), 548–58.
6. L.W. Blank *et al.* (1988), 'New perspectives on forest decline', *Nature*, Vol. 336, No. 6194, 3 November 1988, pp. 27–30.
7. J.L. Innes and R.C. Boswell (1987), *Forest Health Surveys 1987*, Forestry Bulletin 74, UK Forestry Commission.
8. P. Chester (1986), *Acid Lakes in Scandinavia: The Evolution of Understanding*, Central Electricity Research Laboratories, Leatherhead.
9. J. Mulder *et al.* (1990), Effects of air pollutants on man-managed and natural ecosystems, in Stockholm Environment Institute, *Air Pollution in Europe: Environmental Effects, Control Strategies and Policy Options*, Stockholm.
10. UK Acid Waters Review Group, December 1988, *Acidity in UK Fresh Waters*, HMSO, London.
11. J. Ando (1989), Recent developments in SO_2 and NO_x abatement technology for stationary sources, in L.J. Brasser and W.C. Mulder (ed.), *Man and His Ecosystem: Proceedings of the 8th World Clean Air Congress*, Vol. 4, pp. 129–140. Elsevier, Amsterdam.
12. J. Vernon (1989), *Market Impacts of Sulphur Control*, IEACR/18, IEA Coal Research, London.
13. Vernon, ibid.
14. O.R. Young (1989), Science and social institutions, in S. Andresen and W. Ostreng (eds), *International Resource Management*, Belhaven, London.
15. G. Majore, Science and trans-science in standard setting (1984), *Science, Technology and Human Values*, 9 (1), 15–22.

16. D.A. Everest (1990), The provision of expert advice to government on environmental affairs: the role of advisory committees, *Science in Public Affairs*, 4 (1), 17–40.
17. H. Brooks (1984), The resolution of technically intensive public policy disputes, *Science, Technology and Human Values*, op. cit.
18. However, J. McCormick (1989), *Acid Earth*, Earthscan, London, reports complaints that the CEGB was directing research excessively.
19. For the impact of distrust on this communication see K.R. and J.J. Kruper (1989), Communication strategies for resolving environmental issues, *International Journal of Environmental Studies*, 34 (1), 11–23.
20. This is particularly significant in international negotiations where social and political mechanisms for creating consensus are lacking, see S. Boehmer-Christiansen, 1989, The role of science in the international regulation of pollution, in S. Adresen and W. Ostreng, (eds) *International Resource Management*, Belhaven, London.
21. D. Pearce, A. Markandya and E. Barbier (1989), *Blueprint for a Green Economy*, Earthscan, London.
22. Commission of the European Communities, (December 1983), *Proposal for a Directive on the Limitation of Emissions into the Air from Large Combustion plants*, COM 83 (704) final, Brussels.
23. House of Lords Select Committee on the European Communities (1984), *Air pollution*, HL 265, 22nd Report, Session 1983–84, HMSO, London.

PART II
Britain and Germany compared

4 Society, culture and environment

The importance of culture

Cultural differences have undoubtedly influenced the perception of acid rain, the style of response and the measures adopted in the UK and the FR Germany. While culture did not determine acid-rain policy, a more fertile ground for activism was established in Germany in comparison to the situation in Britain.

However, 'moods, feeling, beliefs, values – the stuff of which culture is comprised – can only be studied with great difficulty'.[1] Nevertheless, such 'stuff' is important because it influences perceptions and the degree of emotional involvement of individuals. This in turn affects the amount and types of action individuals are prepared to undertake or approve in the pursuit of specific goals. The observations in this chapter are necessarily of a more subjective nature than elsewhere in this book.

Anglo-German cultural differences have affected the acid-rain debate because they endowed the German debate with a more powerful 'autonomous system of action'.[2] Both air itself and the alleged targets of acid-rain damage had richer symbolic associations, such as freedom, health and general social well-being. Language, which gives useful clues to cultural predispositions, is the focus of some attention in this chapter.

Environmental policy-making does not take place in a vacuum. It is deeply affected not only by objective experience, but also by prevailing concepts of nature[3], attitudes and values which are shared by governments and the wider public. Relevant attitudes include those concerning pollution, forests and the air, as well as the legitimate roles of government and law in regulating the behaviour of industry. Culture appears to have amplified the threat perceptions associated with environmental damage in Germany, while in Britain such perceptions had less impact.

Culture constrains political action and directs it into specific channels. In Germany, it ensured that poor air quality and damage to forests were taken very seriously and aroused emotional responses. In Britain, clean air, apart from specific urban problems, has evoked much less interest and acidification tended to be appreciated only by experts. In addition, the relatively higher status of technology in Germany promoted the more rapid adoption of technical solutions.

How and to what extent public opinion was able to affect the political system in each country is a subject left to later chapters. However, politicians and civil

servants in all societies are themselves members of the public and subject to
wider cultural influences. On the other hand, they may strive to influence
opinion so that government policies gain social acceptance and thus legitimacy.

A word of warning is also warranted. The glib use of concepts such as
'cultural factors' or 'national attitudes' in explanation of policy differences runs
the risk of stereotyping and over-simplification. Cultural perspectives must be
balanced by the more objective analysis of interests, which also underlie
responses to environmental problems.[4] This chapter deals with attitudes and
experience, leaving interest analysis to later chapters.

Values embedded in language and culture

Sensitivity to pollution

Even a cursory observation reveals that the level of orderliness in German
towns and countryside is much greater than in the UK. Standards of tidiness
clearly hold implications for social attitudes towards industrial polluters and
other *Umweltverschmutzer* (those who make the world dirty).

Greater German sensitivity to environmental threats is embedded in language.
In part, this can be traced to the more 'earthy' Anglo-Saxon roots from which
modern German environmental language derives. The English vocabulary for
the environment and pollution control relies on Latin roots which convey a
greater sense of abstraction and distance between people and the world
surrounding them. While the importance of this factor should not be exaggerated,
it may well have played some role in determining attitudes. Three important
linguistic differences with respect to the environment and pollution control can
be identified and include the degree of environmental threat implied by language,
the concept of pollution, and the degree to which solutions to problems are
implicitly guaranteed by the vocabulary of environmental management.

The concept of environment

In English, 'environment' is a broad term the meaning of which depends on its
context. There are distinctions between the human, physical, natural and even
the intellectual environments. The Department of the Environment deals
mainly with housing and local government affairs.

Local authorities have 'environmental health' departments, with the prime
task of inspecting restaurant kitchens. In English schools, 'environmental
studies' are taught as opposed to ecology in German ones. It is appropriate to
talk of the 'British environment' or the environment of a village. The direct
translation of 'environment' into German '*Umwelt*' in the above examples is
quite untenable.

The use of the word 'environment' in its most modern sense is fairly recent
even in Britain. In the 1950s the Central Electricity Generating Board set up an
internal group of experts to advise management on the impacts of power station

siting policy on the environment, i.e. the area immediately outside the factory fence. Environment entered the language of pollution control as a rather narrowly confined term without national, let alone global implications. However, over the last three decades, the application of the concept has widened.

The Germans had to choose a new term for the environment in the late 1960s. *Umgebung*, meaning surroundings, was too narrow and *Lebensraum* ideologically unacceptable. *Naturschutz* (literally nature protection) had other uses. The new word chosen, *Umwelt*, literally means the surrounding world and refers especially to air, water and soil, although its application too has widened in recent years.

Umwelt is not readily applied to small areas or even national parcels. Initially at least, it did not include many of the objects, such as roads and buildings, which the British tend to include in their 'environment'. However much 'nature' may have been altered by human activities, the *Umwelt* remains primarily physical and natural. The range of meanings is therefore narrower than in English, focusing on the natural things which surround us. There is no link with property as there can be in English.

Very importantly, the geographical scope of the *Umwelt* is global rather than national or local. In German, the environment is therefore essentially indivisible; to talk of the German *Umwelt* sounds absurd.

Umweltschutz or pollution control

Environmental pollution control is perhaps the best translation of *Umweltschutz*, literally protection of the environment. *Umweltschutz* is broader than *Naturschutz* and views nature from the perspective of human enjoyment and health. The *Umwelt* is now widely perceived as seriously threatened and heavily burdened (*belastet*) with pollutants and human activities. *Umweltbelastung*, a term widely used to describe pollution and environmental stress, is a concept directly linked to load and weight, i.e. the presence of pollutants. *Umweltschutz* is a matter of state responsibility and public duty.

Schutz itself is a term with a rather wider meaning and usage than 'protection'. It includes guardianship, safeguarding, defence and control among its many meanings. The word also has military uses, including *Grenzschutz* (border control) and *Luftschutz* (air-raid protection). If something requires *Schutz*, the existence of a threat is already implied. Thus, the very term for pollution control embodies a prior assumption that a damaged environment exists and that there is a consequent need to remedy the situation. This function falls upon the state as the defender of society.

The use of the term *Umweltschutz* promises more abatement and raises higher expectations about the outcome of regulation. In English, the terms 'pollution control' and 'environmental management' are more commonly used. While management and control may be achieved by persuasion and good example, for *Umweltschutz*, these measures appear unacceptably weak.

While the term management has been introduced into the German business vocabulary, it has not found a place in law and government, where words meaning 'to lead' or 'to administer' are used instead. The promise 'to manage' is insufficient. The promise of zero-pollution is implied in German usage, but certainly not in English. English tends to be less ambitious, and perhaps rather more realistic, in the degree of redress it promises to pollution victims.

Because *Umweltschutz* implies a national duty and an international responsibility, the Federal bureaucracy in particular had reasons for emphasizing those aspects of German culture which would foster it.

Dirt and pollution

While dirt and pollution are linguistically identical in German, 'pollution' tends to be perceived in Britain as an effect on the environment. Although the term 'pollution' in English holds negative connotations, the corresponding German word, *Verschmutzung*, is stronger and more evocative. The root word, *Schmutz*, means dirt, with further meanings ranging from filth to pornography. Consequently, in German, the distinction between the presence of pollutants and environmental damage is blurred. To the layman they are frequently non-existent. The very presence of dirt tends to imply damage.[5]

This helps to explain the emphasis, in Germany, on emission abatement at source as the primary means for reducing environmental damage. There is less 'conceptual' need to first measure and check whether pollution actually exists. Combined with the high status of the engineering profession and legal/institutional factors described in later chapters, the identification of the very presence of pollutants with environmental damage helps to explain the German preference for technological solutions to environmental problems.

The word corresponding to 'pollutant' *Schadstoff* (damaging or harmful substance), is also more emotive in German because it links damage directly to the substance emitted and hence the polluter. The same word, emission, is used in both German and English to describe the discharge of a pollutant from a source and into the receiving media. However, once in the environment, pollutants become *Immissionen* in German, a word with no English equivalent. *Immissionen* refers both to the presence of pollutants and to their effects. The term most commonly used in the same context in English is 'air quality'. In Germany, while the air is half full of dirt, in Britain it is said to be half-clean. This in turn means that in comparison to English, German tends to intensify and broaden threat perception and the awareness of risk and danger.

Linguistic differences do not, of course, guarantee that German speaking countries will necessarily achieve higher standards of environmental control. However, the German language encourages a high degree of effort, invites more action and promises better results. It may also encourage more disillusionment should such promises not be kept.

Society and the perception of environmental threats

Anxiety and the love of nature

Responding to external threats appears to be a deeply seated human need. At the collective level, external threats are well-known to encourage social cohesion, a phenomenon most widely demonstrated in response to military threats. In the absence of military insecurities, it is possible, however, that environmental threats might play a similar role.[6]

Perhaps for reasons of historical experience, Germans today appear to be more anxiety ridden, and therefore more sensitive to external threats, than are the British. One of the few modern German words to have found a home in the English language is 'angst', with its unique connotations of anxiety and guilt. In the early 1980s, German society seemed to be particularly plagued by angst, one of the consequences of which appears to have been the amplification of environmental threats.[7]

Compared to the British media, the German media pays greater attention to threats, both to personal health and to the planet. Germans almost invariably consider environmental risks to be considerably greater, and action more urgent, than do the British. Erich Wiedemann has rather cruelly summed up German angst as the product of an education system thriving on catastrophe, of *Katastrophenpädagogik*, thus blaming German teachers and philosophers for instilling unwarranted gloom, anxiety and pessimism.[8] He also argues that his countrymen tend to make bad things worse, exaggerate real problems and thus make their solution more difficult.

German forests

The identification of forests as a victim of long-distance air pollution created a powerful image which combined angst with love. This ideal German forest, a symbol of a healthy fatherland and allegedly expressing an erotic as well as neurotic relationships between nature and the *Volk*, has been cruelly satirized in the Republic itself.

The fact that the German forest (*Wald*) became a major symbol in the German crusade for purer air is widely known outside the Federal Republic. Forest death (*Waldsterben*) guaranteed the involvement of many powerful motifs in the acid-rain debate, motifs deeply embedded in culture and literature, ranging from the Grimm fairy tales through to Goethe and the Romantic poets. When during the early 1980s people demonstrated against Waldsterben, they used the slogan 'Erst stirbt der Wald, dann stirbt der Mensch' (first the forest dies, then mankind will perish).

English landscape

German gardens, in contrast to those in Britain, are devoted more to economic production than to the imitation of nature. While Germans may love the idea of forests, many are, in reality, rather ugly plantations which cannot in any aesthetic sense compete with British landscaped parks and gardens. The experience and the idea of nature in the two countries are different. To commune with nature, a German would tend to go where things are free (*ins Freie*) or to where it is green (*ins Grüne*). The British go outdoors, to a garden or to the countryside.

On the other hand, Germans travel in their millions to the Mediterranean Sea where there are no forests. They also deeply admire bare Scottish landscapes. The Germans are indeed a perplexing and contradictory people.[9]

Wealth and insecurity

These observations are borne out by empirical evidence. An opinion survey carried out by the European Commission in 1982, i.e. at the time when major acid-rain decisions were being made, asked people what issues really concerned or worried them. The results, for the UK, for the FR Germany and for the European Community as a whole are shown in Table 4.1. Perhaps illustrating the extent of their angst, Germans expressed, on average, fears about 5.1 issues, compared to 3.6 for Britons.

The survey also revealed very different priorities in the two countries. While in Britain, the three most important fears derived from internal threats (crime and terrorism, unemployment and social tension), in Germany, the main fears, apart from unemployment, related to the despoiling of natural life, artificial living conditions and the impact of chemistry on human health. The considerably lower level of fear in the UK about the despoiling of natural life and 'artificial living conditions' when compared with FR Germany and the European Community as a whole, must therefore be viewed in light of the fact that the UK appears to be one of the least anxious countries in Europe.

Fears for the future are rather different from perceived national priorities. In a 1986 British poll,[10] the environment was ranked as the eighth most important

Table 4.1 Fears of Europeans (in 1982) concerning the future of the world over the next ten to fifteen years

	UK	FR Germany	EEC
Rise in crime and terrorism	77%	57%	71%
Increase in unemployment	61%	75%	66%
Despoiling of natural life	**39%**	**77%**	**57%**
More and more artificial living conditions	**19%**	**62%**	**41%**
Rise in social tension	46%	46%	38%
Critical deterioration in international relations	28%	51%	35%
The risk to human personality in the use of new medical or pharmaceutical discoveries	**20%**	**40%**	**29%**
Invasion of low-priced products from Far East	27%	19%	20%
Others	43%	82%	48%
Sum of all worries and concerns	**360%**	**509%**	**405%**

Note: Environment-related fears are in bold type.

Source: Eurobarometer, 1982.

problem with which the UK Government had to deal, being mentioned by only 8 per cent of survey respondents. The environment was well behind unemployment (75%), health/social services (22%) and law and order (17%). However, views such as these are not static. Environmental issues now enjoy a much higher profile in the UK than was the case in the mid-1980s. In a similar poll conducted in 1989, environment had jumped to the second highest priority (mentioned by 30%), with only health/social services giving rise to greater concern (32%).[11]

Given Germany's repeated physical destruction in the past and its rapid technology-based reconstruction after 1945, 'unnatural' life and artificial living conditions are undoubtedly more widespread in Germany. Germans certainly consume more medicines and have a greater tendency for hypochondria. Taking a cure, a medically recommended holiday in clean mountain air surrounded by forests, is considered to be highly beneficial. It would appear that wealth also encourages insecurity.

A consequence of comparative German insecurity which has some effect on environmental policies may be a greater desire to provide for the future. Not only does this mean that Germans may be more concerned about the loss of national assets, but they also habitually save more of their income than do British people. This links in with wider questions of economic culture and management as will be described in Chapter 7.

In contrast, the British Isles are almost unique in Europe in the degree to which the past has been preserved in the landscape, ranging from medieval villages and towns which are genuinely old, to the winding rural road patterns and bare, deforested uplands and moors. The impact of aggressive agriculture expanding into these areas after having destroyed thousands of miles of beloved hedgerows and ancient copses in the lowlands below, is comparatively recent. It forms a major focus of British environmental concern, and is usually blamed on Europe, i.e. the Common Agricultural Policy.

The British countryside, created by man in a tolerant physical environment, shows few signs of major human failures or blind faith in technology. Rather it is a symbol of Britain's soul and security and a target which acid deposition is not perceived as having damaged significantly.

These Anglo-German contrasts can be explored further by having a closer look at recent history and major geographical differences.

The role of history and geography in threat perception

Culture and values are created by past experience. They filter perceptions of current events. Geography alone has ensured that problems associated with acid rain were objectively different in Germany and Britain. However, history has ensured that these differences were experienced in different cultural contexts. Attempts to develop common measures may therefore be difficult, less because of diverging goals, than because of different initial perceptions.

The recent past

Differences in culture combine with the priorities of political leadership and together influence how slowly or quickly global problems are perceived and how responsive politicians and bureaucracies are to new issues.

In spite of post-war successes in the economic realm, West Germany in the early 1980s, was still less optimistic about the future than post-imperial Britain. The economic miracle itself had left its blots on the landscape. While water and air became cleaner in the experience of many British people after 1945, this was not the case for many Germans.[12] In the FR Germany, much was promised in the early 1970s, but effective implementation proved difficult and slow, not least because of the legalistic nature of the regime and its administrative complexity. In Britain, the promise was less and the perception of slow progress persisted longer. Chapter 9 will deal with these experiences in detail.

The East German example showed West Germans how serious certain environmental problems could become. While this fact was exploited politically, it was also a genuine warning. Sensitivity to air pollution drifting across to West Berlin, and to parts of Bavaria from the industrial south of the Democratic Republic increased during the 1970s as East Germany began switching away from oil to its only significant domestic energy resource, brown coal.[13]

The desire to clean up 'over there' for reasons of self-interest, and as a national duty, should not be underestimated as one among many of the driving forces towards German unification. The response of the energy industries in the much poorer eastern section of a united Germany to these pressures will be of interest and significance in a wider European level.

Commonality and attitudes to Europe

Britain and Germany have very different experiences of European Community membership. Views, while by no means static, differ significantly even today. What was for long the 'Common Market' for the British, has long been *the* European Community to Germans.

The FR Germany was a founding member of the European Community. It helped to create its institutions and generally feels comfortable within them. Political structures and processes in both Brussels and Bonn are characterized by cumbersome consensus politics finding expression in legalistic texts. Also shared is the experience of politics restrained by a written constitution interpreted by a higher court, and the experience of government by bureaucracy rather than elected legislators. Germany remains dedicated to the goal of political integration in Europe.

Britain on the other hand joined late, and only after initial rejection. Legal tradition does not easily accommodate the Treaty of Rome. The British have been and remain divided about the virtues of Community membership, and there is a tendency to attribute recent economic misfortunes to this new and somewhat alien commitment to Continental Europe. EC–UK relations have been abrasive during much of the 1980s.

Many British people do not naturally consider themselves 'Europeans'. If they have strong foreign links, these tend to be with North America and the new Commonwealth rather than Europe. European links are still too often seen as inter-governmental relationships which are not only boring but also suspect in their intentions.

The fundamental relationship with Europe, therefore, has been one of division and defensiveness on the British side, while in Germany many forces combined to establish a more united and positive attitude.

However, attitudes are changing. New forces are emerging in Germany, the environmental movement prominent among them, which are suspicious of the economic and commercial aims of the New Europe. The green movement in Germany tends to see the Community as an institution which will weaken and frustrate its own ambitions. An eastwards shift in the focus of attention could move a united Germany away from the course adopted during the 1980s. In Britain, European integration remains a controversial issue, not least because of high level political scepticism. British environmentalists, however, are increasingly looking to Brussels rather than London for an active environmental policy.

The experience of the national environment

Commitment to the environment was explored by Eurobarometer in 1982 when Europeans were asked 'which . . . ideas or causes . . . are sufficiently worthwhile for you to do something about, even if this might involve some risk or giving up other things'.[14]

This question elicited very little overall difference between the UK, FR Germany and the EC as a whole in terms of total commitment to causes, or commitment to the environment (Table 4.2). While Germans were less concerned with human rights and the freedom of the individual, interest in the environment differed little from that of their British counterparts.

A similar perspective on environmental priorities comes from a 1986 Eurobarometer question about 'what sort of things in life interest you a lot'.[15] Extraordinarily, the environment ranked number one in the UK with almost

Table 4.2   The great causes of Europeans in 1982

	UK	FR Germany	EEC
Peace	58%	57%	67%
Human rights	44%	38%	44%
Freedom of the individual	42%	31%	40%
Struggle against poverty	37%	29%	40%
Protection of the environment	36%	39%	35%
National defence	28%	17%	23%
Others	42%	60%	53%
Total	287%	271%	302%

Source: Eurobarometer, 1982.

half of the people interviewed expressing an interest. The level of interest was marginally higher than in either FR Germany or the EC as a whole. It remains impossible to argue that the British are not interested in their environment. However, the perception of what the environment actually means, what is wrong with it and who should act to improve matters does seem to differ considerably.

Taken together, the survey results demonstrate what any sensitive observer of Britain must suspect, namely that attitudes to the environment as a whole are subtle and by no means as unenthusiastic as might appear from mainland Europe. They are, however, focused differently. The British do attach importance to their environment.

The important question is whether the objective experience of air quality in each country also had an impact on national concern with clean air. The facts of geography are rarely understood sufficiently by politicians; they are difficult to incorporate into legal agreements which prefer uniform solutions and may contradict the aims of economically justified regulation.

Nevertheless, with respect to policy action, the low priority attached to environmental issues in Britain in the early to mid-1980s remains a telling point.

Germany

While history may to a large extent account for more German angst about future environmental damage and loss, geography is also part of the explanation. The meteorological problems of continental interiors differ markedly from those of islands such as Britain, which experience a continuing influx of relatively clean air masses. Germans do not, like the British, live in major conurbations far away from so-called wilderness areas. They inhabit medium-sized and small towns, close to the countryside and woodlands. West Germany is a country of towns, with only Hamburg and Munich, apart from Berlin, being of a size comparable to that of second tier British cities. Consequently, the German population is physically closer to the countryside, and damage to the natural environment caused by pollution may be expected to be more keenly felt.

Access to the countryside is easy, but not necessarily aesthetically rewarding in the Federal Republic. Germany forests are often tedious spruce or pine plantations surrounded by fields without hedgerows. Fewer Germans possess their own garden and access to national parks and large gardens open to the public is more limited. The forested areas in Germany predominate in central and southern Germany, where they tend to become official leisure regions also known as 'pure-air' areas (*Reinluftgebiete*).

Germans witnessed the rapid transformation of nature by industrial technology during recent decades directly, in part thanks to their rapidly increasing personal mobility. It has been shown that stringent emission controls for vehicles would improve air quality for NO_x slightly in Britain by shifting the NO_x 'climate' to the East. However, the effects in Germany, while much more significant in terms of absolute reductions, would not be sufficient to bring air quality up to current British standards.[16]

While forests have a symbolic value in Germany, they are also an important industrial and leisure resource. About one-third of FR Germany is covered with forest and over one-third of this is privately owned. While most forests are exploited for timber, they also raise income through hunting and tourism. They are widely used by urban populations for weekend visits.

For many small farmers and charitable institutions, forests represent a major investment and source of income. Much of the forest is owned by the Länder governments, which have brought considerable political pressure on the Federal Government. For the small farmers and forest owners in the South, who enjoyed political influence during the 1980s, love of the forest was primarily an economic one.

It is part of the German experience to be surrounded by other countries and to be crossed by rivers which have their sources elsewhere. The impact of activities in neighbouring countries encourages sensitivity as well as international interest. The possibility of transfrontier pollution harming German society is readily accepted.

Britain

The awareness of air as an important and threatened aspect of the natural environment is much less intense in Britain. While there is concern and interest in air pollution, there has never been an atmosphere of crisis, as one might expect from a people known not to be easily persuaded or threatened. The British Isles are swept by air which tends to be cleaner than that on the Continent. British people have probably responded to this direct experience by being less concerned about air pollution.

Britain, particularly the prosperous South-East, is a nation of suburbs. While the policy of leaving green belts around major conurbations ensures that people have ready access to non-built up areas, Southern England has no wilderness land. Genuine solitude is virtually impossible without going to the depopulated parts of Britain, such as Scotland. The British also tend to live in one-family houses rather than blocks of flats, leading to the dispersion of air pollution and a reduction in the personal experience of polluted air.

In Britain, air-pollution control became an administrative task, while the public tended its gardens, walked across wind swept moors and strolled through stately homes set in beautiful parks. Like Germans, British holiday makers now migrate to the Mediterranean every summer. Typically, the British spend little time in the more remote areas threatened by acidification.[17]

Political culture and the role of the state

State and society

In Britain, it is sometimes argued, the state reacts to society and the distinction between state and society remains rather blurred. In the Federal Republic, on

Table 4.3   Areas of interest of Europeans

	UK	FR Germany	EEC
Environment	*48%*	*44%*	*46%*
Arts and entertainment	45%	43%	36%
Important social problems	44%	40%	49%
Sport	44%	40%	37%
Third world and underdevelopment	29%	21%	29%
How people live in other parts of Europe	29%	34%	27%
National politics	26%	38%	27%
Science and technology	25%	27%	26%
Regional life, language and culture	18%	19%	18%
International politics	18%	3%	17%
None/don't know	6%	4%	5%
Total	*332%*	*313%*	*317%*

Source: Eurobarometer, 1986.

the other hand, state and society are more distinct. This is reflected in the *Rechtstaat* concept, with the state responsible for leading and improving society through the promulgation of law. Ralf Dahrendorf goes as far as claiming that in Britain 'government in general is still less important ... than in other European countries'.[18]

The concept of the state is powerful and all embracing in German thought, tending to replace that of government.[19] This tends to clash with British ideas of democracy in which public opinion directs policy, rather than being a tool of decisions made by the authorities. It is the role of the German state to protect society. Given a greater sensitivity to threats, the strength of the Rechtstaat concept encourages and justifies action by the German state on issues such as the environment at an earlier stage than would appear to be the case in Britain.

In Britain, during the 1980s, the government has sought to roll back the state even further by weakening bureaucracies and handing functions, such as welfare, back to society and the individual. Individualism and self-help have a longer and stronger tradition in Britain than in Germany. Given the relatively weak powers allocated to the Bonn government after the Second World War, the German Federal Government had incentives to amplify any issue which would strengthen its own role and legitimacy.

British society, and therefore its government, was not receptive to allegations that emissions from Britain were causing damage abroad. Without direct experience of transfrontier pollution it would have been difficult even for a dedicated bureaucracy to create a major public outcry about acid-rain damage.

British acid-rain policy necessarily remained unpoliticized and thus subject to resolution by negotiations between elite groups of top policy-makers and parties with economic interests. The cultural dispositions of these groups would not have encouraged precipitate action, unless powerful regional or national interests had come into play.

State power and national symbols

The German problems arising from regionalism and a relatively weak Federal government are explored in later chapters. They do, however, underline a need for symbols, commitments and issues which encourage national consensus and unity. With German unification until very recently a dormant issue, there had been a need for universal issues which promise to unite right and left. Environmental issues have filled this role admirably since the late 1970s.

In Britain, however, political institutions are highly centralized and there is a wide acceptance of national symbols such as the Monarchy. Whereas patriotic songs such as 'Rule Britannia!' can be sung emotionally and without embarrassment, the situation was very different indeed in the FR Germany. Many songs cannot be sung, verses of the national anthem are avoided and debates about the interpretation of German history and the nature of the German nation continue unabated.

Federal institutions and big business need symbols and policies which involve and represent the whole nation, including the Germans who have been living 'over there' in the Democratic Republic. The need for national symbols and for a national task has therefore been much more deeply felt, and also more difficult to satisfy, in the Federal Republic than in Britain.

The choice of symbols and commitments has also been more limited in Germany because they must not be too politically divisive, i.e. they should neither remind anybody of the era of National Socialism nor be too readily identified with the German Marxist tradition. National symbols should also appeal to the whole nation without arousing fears among Germany's neighbours, by avoiding the impression of emerging nationalism. A strong degree of environmental commitment fits this complex need astonishingly well, allowing Germany to show leadership and initiative in international affairs on a relatively neutral issue.[25] Within a narrowed range of potentially unifying political themes, environmentalism has therefore proved to have a particular cultural resonance for West Germans, reflecting an emerging sense of self-doubt about the sterility of the post-war economic miracle, and expressed largely through the rise of the Green movement.

By way of contrast, in the UK, the sense of loss of international influence and relative economic decline since the end of the Second World War provides an important political undertone. The partly suppressed longing to put the 'great' back in Great Britain is not comforted by grand policies based on idealistic concepts such as environmentalism.

Environmentalism in Britain could be a potent force, but it would have to be based on experience and pragmatism rather than visions of global doom.

It is ironical that, in 1982, both Britain and the FR Germany realized extraordinary degrees of national unity over particular issues. For Britain, national unity came from the Falklands war and its echoing of a lost international role. The German endeavour was a massive technological and economic 'fix' aimed at drastically curbing emissions from power stations, industrial chimneys and car exhausts.

Conclusions

German culture appears to foster a more ambitious attitude to environmental protection than do British values and traditions. This is due in part to the emphasis on cleanliness, but also to a deep seated anxiety, even pessimism, about the future of the world. This may encourage protective state intervention with respect to issues such as the environment at an early stage. The British response to similar questions is more relaxed and a little slower, even if no more reflective.

While the people in both countries are deeply interested in their environment, the concept was understood differently in the 1980s. The British and Germans also have different experiences of air quality in the recent past and different expectations of their governments. Acid rain became an emotional and symbolic issue in West Germany, while in Britain it remained largely a problem to be analysed by experts in consultation with directly affected parties. The ground for the emergence of green politics, explored in the next chapter, was clearly much more fertile in Germany.

Notes

1. R. Withnow *et al.* (1984), *Cultural Analysis*, Routledge, London, p. 3. See also M. Douglas (1985), *Risk Acceptability according to the Social Sciences*, Routledge & Kegan, London. This research strongly supports Douglas's contentions that the neglect of culture is a major weakness in current approaches to the subject and that institutions play a major role in the perception of risk and the potential for environmental damage.
2. Withnow (1984), ibid., p. 1.
3. See, for example, M. Thompson (1986), *Environmental Rationalities*, UK CEED Discussion Paper No. 6, London.
4. For a German view of the British response to air pollution, see H. Weidner (1987), *Clean Air Policy in Britain: Problem-shifting as Best Practicable Means*, WZ Berlin, Sigma Bohn, Berlin.
5. German writers on pollution control generally equate emission reduction with damage prevention, while also regretting the lack of attention paid to scientific analyses, e.g. B. Glaeser (1989), *Umweltpolitik zwischen Reparatur und Vorbeugung*, Westdeutscher Verlag.
6. N. Luhmann (1986), *Ökologische Kommunikation*, Westdeutscher Verlag, Opladen.
7. German distinguishes between two types of threats, active ones involving an identifiable opponent (Androhung), and a more passive one where danger emanates from external circumstances and unpredictable events (Bedrohung). Environmental threats are in the latter category.
8. E. Wiedemann (1988), *Die Ängste der Deutschen: Ein Volk in Moll*, Ullstein, Frankfurt/Main.
9. G.A. Craig (1982), *The Germans*, New American Library, Putnam and Sons, New York.
10. Department of the Environment (1987), Public attitudes to the environment, *Digest of Environmental Protection and Water Statistics*, No. 9, Part 10, HMSO, London.
11. Department of the Environment (1990), Public attitudes to the environment, *Digest of Environmental Protection and Water Statistics*, No. 12, Part 10, HMSO, London.
12. K-G. Wey (1982), *Umweltpolitik in Deutschland*, Westdeutscher Verlag.

13. According to statistics released after November 1989, half the children in the South suffer from respiratory diseases. Forest die-back is worst in the Ore Mountains, where whole ridges are bare due to acknowledged sulphur damage.

14. *Eurobarometer* (June 1982), No. 17, pp. 38–9.

15. *Eurobarometer* (December 1986), No. 26, pp. 36–41.

16. R. Derwent (1987), The motor vehicle contribution to the long-range transport and deposition of acidic nitrogen compounds, *Vehicle Emissions and Their Impact on European Air Quality*, Institute of Mechanical Engineers, London.

17. The disappearance of birds, hedgerows and wetlands, even marine pollution, created more concern and pressure group activity than acid rain. Reafforestation tends to be resisted in the interest of other eco-systems and the panoramic views which rambling through deforested mountain areas affords.

18. R. Dahrendorf (1982), *On Britain*, BBC, London.

19. N. Johnson (1983), *State and Government in the Federal Republic of Germany*, Pergamon, Oxford.

5 Green politics

Introduction

The subject of green politics is of fundamental importance in understanding how energy and environmental policy might become effectively integrated in industrial democracies. This chapter analyses the environmental movements in Britain and Germany and their efforts to influence public and corporate policy through party politics, public opinion and the direct lobbying of decision-making bodies.

In Germany, the party-political system has been significantly altered by green politics. In Britain, on the other hand, the environmental lobby has long enjoyed direct extra-parliamentary links with policy-makers. While the British movement has, in general, been able to make its voice heard, its lobbying on acid rain had little impact on politics and even less on policy. Two questions therefore arise. First, was the more ambitious acid emission abatement programme in Germany adopted because of the different channels of influence available to and used by the green movement? And, second, did the different nature of environmentalism in each country have a significant impact on policy?

The differences between Britain and Germany are to a large extent historically and politically conditioned, reflecting not only culture and tradition, but also the current policy context. Had acid rain emerged as a major issue in the late rather than the early 1980s in Britain, events may well have taken a different course. Green organizations and ideas began to make an impact on the British political system, its operation, goal selection and the resolution of goal conflicts during the latter part of the 1980s. However, environmentalists in Britain largely failed to involve political parties, the public and industry in an open and broad debate about the nature, impact and remedies of acid rain.

In Germany, the situation appears to have been very different. The conclusion reached in this chapter, however, is that German policy did not take the line it did because the Greens forced the major political parties to adopt acid abatement policies against their better judgement. The role of the Greens was more subtle, less direct and altogether more profound in that they modified the political system as a whole and initiated a search for a new policy emphasis.

The acid-abatement policies adopted in the Federal Republic are part of what has been described as 'ecological modernization'[1] or the evolution of a new focus for industrial development dedicated towards a less exploitative relationship between society and nature. Through the European Community,

Britain is now under pressure to acknowledge this attempt to meet, with technological and managerial responses, the challenge posed by the green movement to the established political and industrial system elsewhere in Europe. While the awakening of these responses can in no way be identified with a green utopia, they nevertheless constitute a significant political and industrial change.

Associated with the rise of the green movement in Germany was the growth of a more moderate form of green idealism within the Federal bureaucracy. A reasonable hypothesis is that green politics can enable sympathetic decision-makers to challenge established economic and institutional interests only if party-political competition over environmental protection can be stimulated. Such competition, it has been argued in Germany,[2] is needed to broaden the sphere of action available to the administration in its pursuit of particular environmental objectives. Without green politics, there can be little effective progress in environmental management.

The three components of this chapter are a comparison of the nature of environmentalism in each country, a description of the environmental movement in Britain and a similar account of environmental organizations and their political impacts in the Federal Republic.

Environmentalism in Britain and Germany

Differences in traditions, goals and styles

Four main differences between the roles played by British and German environmental interest groups with respect to energy and environmental policy may be identified. These are the nature and intensity of their concerns, the impact of anti-nuclear protests on public policy, and the influence of electoral systems and party structure on green parties. The last theme is developed further in Chapter 6.

Environmentalism in Germany is largely a post-war phenomenon which emerged in intellectual and political opposition to established structures and ideas, especially the rather single-minded devotion to industrial and commercial achievements.[3] It tends to be radical in its critique of society and fundamentalist in its values. While having deep historical roots in German traditions and experiences, it developed in isolation from existing policy formation processes[4] and therefore lacked the institutional continuity and access to Government which is more typical of Britain. This is a factor which helps to explain major differences in organization and political style.

Truly green goals go well beyond pollution control and the more stringent regulation of emissions and discharges. Distinctions between various forms and degrees of environmentalism are therefore important.[5] The absorption of green rhetoric into the broader political culture is in itself of political significance but does not constitute the adoption of fundamental green goals.

German environmentalists tend to be a darker green than their British counterparts. They emphasize ecology, the frailty of nature and mankind's destructiveness. British environmental groups are more complex in their approach and are less likely to assume the alienation of man from nature. They are readily divided into a well-established and moderate conservation wing, which is seen by the government to have public opinion on its side, and a more recent grouping which takes a more ecological approach.[6] Both sections of the movement tend to eschew political radicalism, and the conservation groups in particular do not present any challenges to fundamental social structures and values.

The British concept of conservation covers not only nature but also the cultural heritage. The German translation of conservation, 'nature protection', is thus very imperfect. Typical of British conservation groups has been their relative disregard of industrial pollution in favour of countryside issues, yet another concept which cannot be translated into German. The recent history of environmentalism in Britain is therefore characterized by less conflict and a low level of party-political competition over environmental policy.

Neither was the environment a party-political issue in the Federal Republic until the mid-1970s. Change came about when a growing grass-roots movement clamouring for influence on policy, which had promised much but had delivered little, was largely excluded from rather rigid decision-making structures.[7] Planning controls in particular were weaker in Germany than in Britain. Forcing their way into the planning processes was accompanied by the politicization of the participants and a radicalization of values.

In contrast, British environmental groups have traditionally made use of their access to policy-making circles inside the administration. Direct lobbying of the government and Parliament (and, more recently, the institutions of the European Community) offers a much more likely route for influencing policy. However, they have also remained considerably weaker in their ability to affect policy than their German counterparts.

One major difference in the goals of environmentalists in each country was the attitude and attention devoted to the nuclear issue. While German environmentalism largely grew and organized itself in opposition to nuclear power stations, radioactive waste disposal and reprocessing,[8] overall responses were more muted and diffuse in Britain. The links between nuclear policy, official energy programmes and environmental policy were and remain more explicit in Germany, but hardly existed in Britain. In Germany there were clear political links between acid rain and nuclear policy because the coal and nuclear options for electricity generation were hotly debated during the late 1970s (see Chapter 8).

In contrast, British environmentalists could rely neither on the politicization of energy policy to draw attention to their concerns, nor did they have much support for acid emission abatement from within the administration. Policy-making was thus made in a more technocratic fashion by relatively few people.[9] In Germany there was so much agitation about the environment, and especially the threat of air pollution, that policy might appear to have been made in the streets and by the media.

British pragmatism and elitism

In Britain, continuity, philosophical pragmatism and attention to local issues of amenity characterize both the old and, to a lesser extent, the newer environmental groups. In a sense, British environmentalism, like society at large, is organized hierarchically in that groups, domiciled in London with strong leaders, decide national policy and lobbying strategy. For example, Jonathon Porritt of Friends of the Earth (FoE)[10] has dominated the environmental conscience of Britain for over a decade to an extent which no similar individual was able to do in Germany.

Green arguments and concepts were of course debated in Britain during the early 1970s at the time of the limits to growth debate. The Green Party (known as the Ecology Party prior to 1985) was formed at that time. Yet lack of public support for its deep green views,[11] combined with the first-past-the-post electoral system, made its failure to have any direct impact on party politics a foregone conclusion. It is virtually impossible for new groups to break into British party politics as a result of the dominance of the two main political parties. Even the Green Party's extraordinary 15 per cent share of the vote in the 1989 European elections did not result in representation in Strasbourg.

Instead, environmentally concerned activists and voters moved towards the smaller parties in the political centre, formed pressure groups within existing parties, or set up new environmental lobbies which tried to influence the government and public opinion in a non-political way. None of these groups presented significant threats to any one party's political power.

While both countries imported environmental ideas from the United States during the early 1970s, in Britain these tended to be either criticized or ignored. On the other hand, German culture and especially potential critics of the status quo were much more receptive to these new ideas. In Germany, the Club of Rome's *Limits to Growth* became a best-selling paperback and had an impact on government policy.

Warnings from British environmentalists are in general less intense and less overtly critical of social and political structures. This is perhaps not surprising because Britain was approaching a zero-growth economy during the mid-1970s. Environmental activists operated in an intellectual atmosphere of gloom about the country's inability to remain competitive in world markets and to advance technologically as fast as either Japan or West Germany.

In the 1970s British political culture was still based on consensus. Evolutionary change through reasoned argument and established channels was the way forward for the environmental lobby. Given the conservative nature of British institutions and the failure of previous challenges to the two-party system, British environmentalists had little hope of success through a head-on onslaught on the political system. The more fruitful route for any citizen hoping for more demanding environmental policies was through the traditional British environmental lobby.

Environmental politics have tended to be confined either to local issues, which local MPs can and do take up effectively, or to lobbying on larger issues which remain very much a matter for government unconstrained by party-political debate.

Green politics, in the sense of party competition, largely post-dates the acid-rain debate in Britain. Being green (in a relatively weak sense) gained considerably in respectability after the Prime Minister, in a speech to the Royal Society in 1988, warned about mankind's impact on the environment. However, this rhetorical re-orientation of Government policy took place not in relation to acid rain, but to ozone depletion and climate change. It is, however, possible that Mrs Thatcher's 1988 speech would not have been politically credible had Britain not, a few months before, made concessions on acid-rain abatement to its EC partners.

The issue of party-political responses to acid rain has been weak enough in Britain to be ignored for the purpose of analysing policy. The main focus will be on the impact of environmental groups on the decision-making process, their respectability and, in comparison to Germany, their relative tameness.

Ecological fundamentalism in Germany

In West Germany, environmentalism as social critique and protest had both more justification and opportunity to express itself and thus to effect the legal and the political system. While both German and British greens can be described as ecological fundamentalists searching for a new society, the German greens have had a greater influence on the environmental movement as a whole. Regional parties were created which eventually coalesced to enter the Bundestag (Federal Parliament) as the Die Grünen in 1983.

The adoption of very stringent standards of environmental protection, whenever technology permits this, is by no means the German environmentalists' dream. At best it is a minimum adjustment through technological fixes, at worst a travesty serving to stimulate rather than restrain personal consumption. To the Greens, technological fixes may further widen the gap between rich and poor throughout the world and promote industrial concentration and political centralization.

Technological fixes, such as flue gas desulphurization and catalytic converters, while they cannot, for pragmatic political reasons, be opposed by Greens, are not the primary objectives of ecological thinking. The true goals are reduced consumption, energy conservation, recycling of material, the reduced production of toxic waste and, above all, environmentally responsible behaviour at the individual and social level.

As in Britain, environmental programmes (including major legislation on the statue book) were generally not well implemented during the 1970s. This was due to continuing political concern about economic impacts. Unlike their British counterparts, however, the German Greens had a much greater chance of making a political impact in the 1980s because of the system of proportional representation used to elect members of the Bundestag. They were also able to frighten government seriously, especially about the future of nuclear power, an issue on which they had widespread public support.

Political culture, the real experience of industrial growth and environmental damage, as well as the interests of some senior administrators and politicians assisted them further. The Greens could raise the political salience of environmental destruction sufficiently to make pollution abatement a matter for party political competition and administrative activism. From the very start, German environmentalists could rely, in certain areas, on the backing of the law and implicit support from new and dedicated sections of the Federal administration.

The environmental movement in Britain

Origins and actors

In Britain, environmental and conservation groups have consciously avoided participation in party politics. Criticism of the environmental records of the mainstream political parties has remained studiously even-handed. There has been little support from environmental groups for the fledgling Green Party which is perceived as a dubious vehicle for realizing environmental ambitions. As in Germany, only more so, membership of environmental groups has been highest among the middle classes.

The network of autonomous groups which forms the British environmental movement also builds securely on a strong tradition of British voluntarism and provides important bridges between local concerns about amenity and conservation and larger-scale national or regional environmental issues.

The different environmental groups have also become practised in co-ordinating campaigns and informally sharing out issues between themselves to avoid duplication of effort. In a practical sense, the environmental movement is a great deal more than the sum of its component parts.

The well-established conservation groups, the activities of which have revolved around the concepts of stewardship and national heritage, often promote the direct interests of their members. The Rambling Association for example promotes walking access to moors and coastal paths while the Royal Society for the Protection of Birds (RSPB) encourages access to and protection for areas attracting birds.

Compared to the German green movement, the more traditional British groups have in many ways remained closer to nature as experienced directly by their members. Campaigns have remained down-to-earth and were directed not at the world, but at local surrounding. Less tinged with utopian ideals and emotional protest, traditional British groups were arguably less able on a cultural level, and institutionally ill prepared, to have an impact on the large-scale international issues of the 1970s.

On the other hand, there is the more recent wave of environmental groups of which FoE and Greenpeace stand out. The conservation and ecological groups are distinguished in a number of ways besides the length of their history. There are differences in internal structure, mode of operation and relationships with government and the public.

Environmental groups – the first wave

The groups making up the British environmental movement are the result of three historical waves of interest in nature and ecology.[12] The first, in the late nineteenth century, gave rise to groups such as the National Trust, the Town and Country Planning Association (TCPA) and the National Society for Clean Air (NSCA). The interests (and names) of these groups have evolved considerably during the last century.

The National Trust, founded in 1895, has 1.8 million members and owns 450 000 acres of countryside as well as miles of coastline and hundreds of stately homes, making it the UK's largest private landowner. It has more members than all the British political parties combined. The Trust perhaps best expresses the British public's attitude to the environment – love of the countryside and the national heritage as represented in buildings of historic interest. It symbolizes the enormous value placed in Britain on the leisure opportunities the countryside is seen to offer.

The National Trust is identified as being at the core of the British conservation movement. Although a private charity, the Trust has a special position because a 1907 Act of Parliament enabled it to declare its land and buildings 'inalienable' – i.e. protected from sale or mortgage. Much of its inalienable land is also protected against compulsory purchase by the right of appeal directly to Parliament.

While the National Trust may have some superficial similarities to the German BUND described below, it functions very differently. It can in no way be seen as a pressure group which challenges the political establishment. The National Trust is very much part of the British establishment and has played a critical role in reconciling the desire of the aristocratic classes to maintain their ancestral homes with that of the wider public to experience at first hand their national heritage.

There have been some tensions within the Trust concerning the relative weight which it should give to preserving landscape as opposed to buildings and architectural features. The Trust was not involved in the acid-rain issue.

The TCPA was established in 1899 as the Garden City Association. While its central interest remains the promotion of enlightened planning, it has latterly shown a great interest in energy issues and has participated in a number of recent public inquiries, including those concerning the Windscale nuclear waste reprocessing plant and the Sizewell nuclear reactor. Given its particular interest in planning issues, the TCPA has not been active in the acid-rain debate.

The NSCA began life as the Coal Smoke Abatement Society in 1899, changing its name to the National Smoke Abatement Society in 1929, and becoming the National Society for Clean Air in 1958. Its objective is to 'secure environmental improvement through the reduction of air pollution, noise and other contaminants'. Local authorities are the mainstay of the NSCA's membership, although many major industrial concerns, including the Central Electricity Generating Board (CEGB), British Gas, British Coal and several oil companies are sustaining members.

From its Victorian origins, the NSCA has matured as far as is possible for an environmental group. A grant from the Department of the Environment enables a Library and Information Service which is made available to the public. The NSCA has built up an extraordinary degree of professional expertise through its membership, allowing it to be consulted by government and to submit evidence to parliamentary inquiries. The NSCA's heyday was the immediate post-war period, the Society being responsible for drafting much of the 1956 and 1968 Clean Air Acts.

In the early 1980s, the NSCA came out strongly in favour of a more ambitious UK policy to combat acid rain. On the other hand, it did not, until 1989, support the introduction of catalytic convertors on cars on the grounds that the scientific evidence was not sufficiently convincing.

The 1920s wave

Following the end of the First World War, a second wave of interest in nature and conservation issues occurred. The most prominent of the groups to have been established at that time is the Council for the Protection of Rural England (CPRE), which was formed in 1926 through the amalgamation of a number of other national and local groups. Until after the Second World War, its main purpose was to promote legislation on countryside issues, particularly on national parks and green belts round urban areas. After seeing many of its major objectives embodied in the 1947 Town and Country Planning Act, CPRE's national role became less clear. However, it has experienced a renaissance since the 1970s, when it began to focus on the issue of development in rural areas.

CPRE has devoted considerable attention to the need for and selection of power station sites. It has also questioned the need for investment in new coal mines. CPRE has come to be regarded as one of the most important of the non-local objectors at energy-related public inquiries. It took an active part in both the Sizewell and Hinkley Point Public Inquiries. CPRE lobbied vigorously for the inclusion of clauses imposing statutory obligations to protect the environment and promote energy conservation in the 1989 Electricity Act privatizing the electricity supply industry. The imposition of environmental duties on electricity generators has developed into a touchstone of faith for CPRE which was instrumental in securing the inclusion of the relevant clauses in the 1957 Electricity Act.

CPRE also fought extremely hard against those components of the 1989 Electricity Act which the government had intended using to support the further development of nuclear power. It commissioned a legal opinion on the compatibility of the government's 'non-fossil fuel obligation' with European competition law.

The CPRE's publication, *Countryside Campaigner*, today reaches more than 44 000 members, many of whom are influential members of the community, at both local and national levels. CPRE, once again, plays a key role within the UK environmental movement and is the epitome of the more conservative forms of British greenery, standing for the protection of a rural environment which is seen to represent the core of English culture. CPRE has increasingly been ready to form alliances with other environmental groups such as FoE or WWF.

Environmental groups in the 1970s

The 1970s saw the emergence of a set of new pressure groups, very different in style and mode of operation from the traditional conservation groups which had operated up till that time. The more radical style of groups such as Greenpeace and FoE has attracted younger members, with the result that these groups might too easily be seen as encompassing the entire British environmental movement.

The new groups differ from the older conservation groups in three key ways. First, the public tend to be enlisted as supporters rather than members. As a result, policy is made by the group's leadership and its officers without the use of cumbersome consultation procedures. To some extent, this merely reflects the effective practice of some of the longer-established conservation groups with more elaborate constitutional procedures.

A second difference is that the newer groups are motivated by a more ideological attitude to the environment. Leading on from this, a third difference is that the modern ecology groups leave open the option of taking direct action, i.e. backing up their beliefs with methods which go beyond the law. Blocking discharge pipes at Sellafield and obstructing whaling fleets are obvious examples of direct action.

As a result, the modern environmental groups cannot merely be seen as pressure groups which represent the interests of particular groups of people or the public at large. They are not just channels for public concern, but play a much more active role.

First, almost like companies developing a new product, they identify those issues to which their membership and the public may be most receptive and 'market' solutions through professionally organized campaigns. The second function is to put pressure on government and industry to respond to public concern and adjust policies so as to remedy perceived problems. The success of an environmental group is a measure of the degree of entrepreneurship which it shows in identifying problems and 'selling' solutions. Whether it was intended or not, Mrs Thatcher's Royal Society speech made the task of environmental groups easier and supporters' numbers having grown dramatically.

The staff of the larger environmental groups have come to display a degree of professional skill which matches that found in government and industrial organizations. As a result, the more mature environmental groups tend to enjoy regular access to Parliament and government through personal contacts and through membership of working groups and consultative committees. Thus, the pressures exerted by environmental groups derive institutional momentum from their skill base and networks of contacts.

Friends of the Earth and Greenpeace

FoE was the first of the modern environmental groups to make a large impact in the UK. Like the National Trust, it has been identified, in a very different way, as being at the heart of the British environmental movement. FoE's aim is 'to seek the prevention rather than the cure of environmental destruction and thus to concentrate efforts on policy changes rather than on specific cases of environmental abuse'.

It generated a great deal of its initial support in the early 1970s by focusing on the whaling issue which had a strong emotional appeal. In the early days, FoE also used direct action, though of a somewhat tamer variety than that used by Greenpeace, with its famous organized dumping of Schweppes bottles on the doorsteps of the company's headquarters. However, FoE now works within the law, aspires to having greater direct influence on Whitehall and commissions consultant's reports, which are subsequently refereed, to provide a factual underpinning for its campaigns.

FoE regularly submits evidence to parliamentary inquiries and is often called in to give oral evidence. FoE has used this avenue to voice its support for measures to abate acid rain since the early 1980s. In collaboration with other organizations, FoE has hosted conferences on clean coal technology (with British Coal and the International Coal Development Institute), forest die-back (with the Goethe Institute) and vehicle emissions (with vehicle manufacturers).

In 1984, the FoE annual conference was addressed by the junior environment minister, William Waldegrave, symbolizing more than anything else FoE's emergence as a mainstream, almost 'establishment' environment group. FoE is also consulted and briefed by the government on a more informal basis. This may demonstrate the gradual emergence of a new informal alliance, not unlike the one which developed between the Federal Government and the Federation of Citizens Groups for Environmental Protection in Germany in the 1970s.

There are now about 200 000 FoE supporters compared to 26 000 in the mid 1980s. However, FoE's movement into the suburbs of British policy-making has lost it the support of some of its more radical members to other groups, such as Greenpeace, which have been more prepared to go outside the law. FoE is well advanced on the progressive path from being an idealistic fringe group to a mature lobbying organization. Professionalism is replacing enthusiasm as the prime qualification for its staff. However, in spite of FoE's London base, much of its strength remains at the grassroots in its powerful network of 220 autonomous local branches.

British Greenpeace was founded in 1976. Greenpeace has made itself one of the most attractive environmental groups to the young through its policy of hitting the headlines with direct action, such as interfering with whaling fleets and sailing its ship, the *Rainbow Warrior*, into the French nuclear testing area in the South Pacific.

Until recently, Greenpeace has made relatively little use of formal procedures such as public or parliamentary inquiries to promote its causes. It has been the least trusted, by government and industry, of the new generation of environmental groups and has, on occasions, been hyperbolically referred to as a 'terrorist' group because of its direct action. In the past, government officials have declined to address meetings organized by Greenpeace.

However, following FoE, Greenpeace is beginning to move towards a position of more general acceptability. It hosted an international symposium on acid rain during International Acid Rain Week in early 1987. It is proposing to organize a conference on acid rain jointly with the British Library in 1991 and has submitted evidence to parliamentary inquiries on acid rain as well as government consultation papers on acid-emission abatement proposals. Greenpeace also commissions research from external consultants to underpin its campaigns.

Greenpeace and FoE are known to liaise with each other, ensuring that they do not duplicate campaigning effort. When in 1989 the environmental movement attacked car manufacturers for failing to install catalytic convertors on new vehicles, Greenpeace and FoE each targeted a separate company resulting in a co-ordinated and effective campaign.

Other modern groups

Traditionally, the World Wide Fund for Nature (WWF – formerly the World Wildlife Fund) has focused on nature conservation issues, symbolized by its familiar panda logo. However, at the same time that it changed its name, WWF has begun to develop a sharper campaigning edge on issues going beyond the boundaries of nature conservation. Starting from a privileged position and being relatively well-off because of corporate sponsorship, WWF has developed its influence very rapidly. It continues to attract large sums of money from sponsors such as the National Westminster Bank. The Central Electricity Generating Board also supported it. WWF makes funds available for commissioned research by other environmental groups such as FoE or CPRE. It thus acts as a channel for the indirect corporate funding of environmental groups which might otherwise find it difficult to acquire money from that source.

The much less visible Green Alliance is a club of about 420 academics, industrialists and environmentalists 'distinguished in their own careers by their concern for the environment'. The purpose of the exclusivity is to ensure that it 'shall neither have more than its fair share of nuts nor be liable to takeover by extremist groups'. The Green Alliance is, therefore, through its discretion and self-selection, an environmental pressure group ideally suited to influencing policy within the British system.

The aims of the Green Alliance are to build a substantial constituency in each political party to promote an ecological perspective; to further understanding of political processes among groups and individuals concerned with ecology; and to further the development of 'green' ideas. It was founded in 1978 by Tom Burke, who was executive director of FoE for five years. Given the experience and the contacts of its members, it can have an important, though somewhat invisible, influence on British environmental policy-making. There is a 'Friends of the Green Alliance' organization which provides wider support.

The British Green Party

The British Green Party was founded as the Ecology Party in 1972 by 'deep' ecologists. It changed its name in 1985, adopting at the same time the sun flower symbol used by the German greens. The Green Party's usefulness has recently been defined by Tom Burke as being limited to its ability to put pressure on other parties. He believes that its stated aim of political power is unrealistic. Nine-tenths of leading British environmentalists apparently did not support the Green Party in the 1989 European election.

As for many German greens, the Green Party emphasizes a spiritual and almost religious dimension which at times verges on the anti-rational. The Party supports British withdrawal from NATO, unilateral disarmament and is strongly opposed to nuclear power. Improved efficiency is the basis of its energy policy and thus comes closest to the German Green Party.

The British greens astonished political commentators when they polled 15 per cent of the vote in the 1989 European elections, outpolling the traditional centre parties. However, this strong showing did not result in any representation in the European Parliament. The high green vote has been attributed to a protest vote by disgruntled government supporters at what was seen as a non-crucial election and to the movement of supporters of unilateral disarmament from the Labour Party. However, a widespread feeling that the government and opposition parties had not taken a strong enough position on environmental issues must also explain the high green vote.

The Greens have now settled down to about a 5 per cent rating in opinion polls, probably representing a core of support beyond which the party will not progress. Nevertheless, the Greens' ephemeral success was vital in maintaining interest in environmental issues in the UK within a year of Mrs Thatcher's Royal Society speech. The mainstream parties are now adopting some policies and postures acquired from the environmental movement.

The British environmental movement and nuclear power

Whereas German opposition to nuclear power often took the form of mass protests, the respons of British environmental groups has been more sedate. Environmental concern and nuclear protests did not converge to nearly the same degree as they did in Germany. This partly reflects Britain's role as a nuclear weapon state. Public protest tended to be directed against nuclear weapons, drawing attention away from the civil nuclear power programme. Anti-nuclear weapons protests were organized by the older mass-support organization, the Campaign for Nuclear Disarmament (CND), which, after years of failing to change defence policy, has only very recently moved in a green direction.

A German perception of British opposition to nuclear power is that it has been 'isolated and of little significance'.[13] However, British opponents have had other channels available to them through the public inquiry system. Four major nuclear-related inquiries have taken place in the UK since the late 1970s, covering waste reprocessing at Windscale (renamed Sellafield) and Dounreay, and the construction of US-designed nuclear power stations at Sizewell and Hinkley Point. The mode of opposition has been closely argued and documented legal submissions rather than emotional public protests.

In 1989, the government abandoned plans to force the development of more nuclear power in the UK. This appears to have been the consequence of a more realistic appraisal of the economics of the technology taken in advance of electricity privatization (see Chapter 8). However, responses to safety concerns expressed by environmental objectors at public inquiries have undoubtedly forced up the price of nuclear power and may have contributed indirectly to the Government's decision.

Environmental groups such as FoE, Greenpeace and CPRE have been particularly active in opposing nuclear power through public inquiries. However, many well-established groups have taken little interest in the nuclear issue.

The environmental movement and acid-rain policy

While several environmental groups have been concerned with the acid-rain issue, few have been very active. The environmental groups have not succeeded in shifting British acid-rain policy to any large extent.

FoE was a major participant in the Stop Acid Rain Campaign, led by Scandinavian environmental groups, which was active in 1984/85. This campaign focused on the broader global impacts of acid rain and the more ethical dimensions. Although car bumper stickers blossomed for a while, the UK campaign fizzled out. Both FoE and Greenpeace participated in International Acid Rain Week, held during May 1987. FoE organized a cycling rally in London, while Greenpeace hosted an international symposium. Both FoE and Greenpeace have been submitting evidence to appropriate parliamentary inquiries.

In 1985, FoE published a commissioned report covering all aspects of the acid-rain problem ranging from abatement costs through to environmental damage in Scandinavia.[14] This report was a more sober account of the acid-rain problem than many of the more lurid accounts which appeared on the bookstalls in the mid 1980s. The report raised the question of the effects of acid rain on freshwaters and forests in parts of the UK. It received good press coverage when it was released at a high point of political interest in the issue.

FoE also submitted evidence to a 1984 House of Commons Inquiry into acid rain[15] and was called in to give oral evidence. Its policy recommendations were later echoed by the committee.

In attempting to raise public anxiety, FoE focused a great deal of attention on the issue of potential UK forest damage from acid rain and the perceived complacency of the Forestry Commission. FoE commissioned a survey of the health of British trees and invited West German forest experts to assess the UK situation.[16] Although FoE's scientific methods were criticized because of inconsistency, the survey achieved the objective of putting the Forestry Commission on the defensive and ensured that official bodies made more effort to assess systematically the health of UK forests.

While FoE's forest surveys received a reasonable degree of press attention, they did not ignite public opinion. In this sense, forest damage was probably the wrong topic on which to focus, as trees, least of all conifers, are not at the heart of the British love of the countryside. Indeed, recent press attention on forests has been negative, focusing on the tax benefits obtained by prominent public personalities investing in softwood plantations which are seen to disfigure the landscape.

Since 1984 many of the established environmental groups have taken a pro-active position on the acid rain issue and lobbied government through more discrete channels. Notably, the NSCA has articulated a formal policy on acid rain (calling for tighter controls) and co-organized a high-level conference on the theme of 'Acid Rain: The Political Challenge' with speakers from the UK government, the CEGB and various European institutions.[17]

The environmental movement in Germany

Citizen action groups and nature protection

Why did a highly political movement arise in Germany, when similar attempts in Britain, where environmentalism was in many ways more securely established, failed? Why did a band of rather diverse 'drop-outs' climb to the top? Only history, culture and difference in the political system can suggest answers.[18]

The Green Party drew its main support from several strands of the wider, and less ideologically inclined environmental movement. Among the most important were the Bürgerinitiativen (citizen action groups) which mushroomed in the Federal Republic during the 1970s. Almost half of these groups came to concern themselves with environmental issues.

The Bürgerinitiativen had their roots in towns and cities and, initially at least, sought participation in local planning decisions, a right which British citizens already possessed to a much larger degree. The Bürgerinitiativen had become politicized during numerous anti-nuclear and nature conservation battles fought during the 1960s and early 1970s. Battles against planning authorities brought only occasional and limited success and the conclusion was drawn that established political and scientific structures (rather than science or politics as such) were their chief adversaries, a view from which the Greens themselves came to benefit.

The Bürgerinitiativen began organizing themselves federally in 1972 by founding the BBU, the Federation of Citizens' Groups for Environmental Protection. These beginnings, it is widely believed in Germany, were actively supported by the newly set-up Federal environmental administration. The BBU came to be the major grass-roots support organization of the Green Party.

By 1980, there were over 130 supraregional and over 1100 regionally active environmental groups in West Germany, although not all had joined the BBU.

The BBU organized itself alongside the somewhat older BUND, the German Federation of Nature Protection. The BUND had developed from the Deutsche Naturschutzring which was founded in 1950 and which had proposed, during the sixties, a Green Charter for nature conservation and landscape protection. None of these organizations would describe themselves as non-political in the German sense of the word, although certainly as independent of any political party.

In contrast to the BBU, the BUND tends to be backed by the owners of large forests, including members of the aristocracy with considerable political influence in some Länder. In many ways, the BUND remains a deeply conservative organization, but cannot be equated with either the National Trust or the CPRE. It is more likely to defend the perceived interests of rural property owners, as opposed to cultural heritage as such. Its influence has been particularly great in Bavaria and one suspects that many members of the BUND would vote green only when the conservative CSU could not deliver.

Science and the media

The BBU in particular began to use science, the media, paperbacks and academic legal arguments to advocate 'soft technologies' and challenge established institutions, and through them, industry and developers. It found support and help from experts on scientific and legal matters, such as the Freiburg Institute for Applied Ecology. However, the BBU had difficulty in doing this effectively, because of its non-hierarchical and intentionally loose organizational structure. It therefore directed its messages towards the general public rather than to politicians and administrators.

The BBU believed deeply in converting the grass roots, and has been described as the organized educated middle class dissatisfied by the established party system and its values. Social circumstances encouraged the green movement and the rise of the Green Party to political influence. In 1975 eighteen year olds were given the right to vote, and it was the 18–29-year-old age group from which the Green Party, led mainly by forty year olds, came to attract most support.

One function of the green movement has been to act as a safety valve for some of the violent impulses and frustrations which disturbed German society during the 1970s. The green movement thus brought together forces of dissent, critique and disillusionment from along the whole spectrum of German ideology and political belief, united by little more than the desire for non-violent reform and faith in 'ecology'.

Green politics and nuclear power

The central objective of the green movement during the 1970s was to rid the Federal Republic and the world of nuclear power. Green politics thrived therefore on public controversy over energy policy. To pursue this objective, it either had to make an impact on the political system or use the Länder administrative courts. Both opportunities were taken up.

Anti-nuclearism in Germany has a long intellectual history and has been supported by philosophers of both the right and left (e.g. Heidegger and Fromm). Mankind and culture were seen by both as threatened by technological gigantism and irresponsible consumerism. The struggle against the state-sponsored nuclear programme was widely interpreted at the time as part of the post-industrial protest movement. This challenged not only the concept of large scale technologies as the engine of progress, but also capitalism as the highest form of economic organization and, particularly relevant in the German context, the idea of the benign state as the leading institution in society. Fears that the Government might want to develop nuclear weapons played a significant role.[19]

Not all of the first commercial nuclear plants built in the 1960s attracted significant local opposition. The planned expansion of nuclear power plant construction after the 1973 oil crisis, however, encouraged a sharp increase in protests and thus the politicization of energy policy in general.

Towards the end of the 1970s the economic situation in Germany worsened. Anti-nuclear demonstrations grew more violent and industry tended to blame the Greens for both. The first serious protests took place at Whyl, near Freiburg in Baden-Württemberg. 26 000 people demonstrated against a nuclear power station to be built at Brokdorf near Hamburg in the autumn of 1976, an event which established a pattern of escalating violence. Brokdorf attracted 100 000 protesters in early 1981 and the Kalkar fast breeder reactor 30 000 in October 1982. A press campaign allegedly took place against nuclear opponents in Bavaria in 1983 and academics felt that public opposition to nuclear power began to endanger their jobs.[20] A year or so later, when construction work began at the Wackersdorf reprocessing site, the Bavarian police was equipped with CS gas.

Links with the peace movement

In the early 1980s a resurgent and rapidly growing peace movement blossomed in response to NATO's nuclear modernization proposals and the promised arrival of new Cruise and Pershing missiles. This deeply divided German society and further inflamed anti-nuclear feelings. Links between the Greens and the peace movement grew. There was a jointly organized protest of one million people against rearmament and the visit of US President Reagan in Bonn in the summer of 1982. The fear that there was a genuine link between civil reactors and the desire for nuclear weapon status by Germany grew stronger.

In the end, the two movements did not merge because of fundamental disagreements between pro-communist groups, which figured largely in the peace movement, and the libertarian and decentralizing Greens. Nevertheless, a green/peace movement alliance appeared a real and destabilizing possibility to many Germans in the early 1980s. It will later be shown how acid-emission abatement policy was one factor which helped to defuse that situation.

The nature of green protest

West Germany is considered, in Britain, to be a stable, wealthy and enviable Western democracy which owes its success to the civil application of science and technology. There is a different side to this picture, especially for younger people afflicted by anxiety and fear about the future of industrial society and its wasteful if enjoyable habits.[21] High levels of consumption were seen to conflict increasingly with other strands in German culture such as romanticism, 'technological hostility' (Technologie-feindlichkeit), Marxism and Christian self-discipline. Economic success does not satisfy German culture and there is a genuine search for some utopia or unifying idea.

By 1979, ecologically concerned people had joined forces with the remnants of past opposition movements. The politicized youth of the sixties, organized into the extra parliamentary opposition (APO), had largely failed to reform society. It was to provide many of the older leaders of the green movement and later the Green Party. These two groups were supported by a miscellaneous collection of vegetarians and worshippers of anthroposophical philosophies, as well as younger people who had either opted out into alternative lifestyles, or whose social and economic success was accompanied by feelings of guilt.

Many of these young people felt resentful of the State and its established hierarchies and hypocrisies which made entry and change difficult. They challenged the legitimacy of the political status quo, though less violently than the Bader-Meinhoff Gang before them.

As the economic situation worsened during the late seventies and unemployment grew, green ideas about linking environmental protection with the preservation of human dignity and job creation became more appealing. This brought some new supporters from the trade union movement. However, the trade unions and the Green Party did not begin to talk formally to each other until 1985.

A deeply felt score which the Green opposition felt it needed to settle with the older generation was their silence about the National Socialist era. Green speakers in Parliament were the first Germans to express publicly, often in a self-righteous way, shame, hatred, and contempt for those who had collaborated with and benefited from the Nazi regime and who now appeared to value little else but material success. From the perspective of the conservatives, greens were seen as disloyal Germans intent on 'dirtying their own nest'.

A broader intellectual attraction of environmentalism lay in its claim to be both systematic and holistic, analytical as well as integrative. It allowed a certain convergence between Marxist thought and conservative ideas about man's duties to God and society. It was international in theme but did not conflict with love of Heimat (the homeland).

The green activists combined relatively new ideas about environmental catastrophe and global over-exploitation with more native traditions of idealism and utopianism, attractive to individuals plagued by feeling of guilt and longing for environmental wholeness, for landscapes with fewer autobahns, power stations and ski slopes.

The German Green Party

Calling themselves green, to distinguish themselves from the SPD and Communists (red), the CDU/CSU (black) and the FDP (blue), green activists learnt their political skills at the communal and regional level during the mid-1970s. In 1978 almost 60 per cent of voters in a national opinion survey expressed sympathy for green groups. In order to enter German parliaments, a party needs to obtain only 5 per cent of the vote. The new party was clearly a threat to existing political interests.

The first Green Party, the Grüne Aktion Zukunft (GAZ), was founded in July 1978 by a lapsed CDU member of the Bundestag, Herbert Gruhl, who accused his party of having raised the traditional sins of avarice, envy and greed to moral virtues, and of having turned homo sapiens into homo economicus. In 1979, green coalitions participated in several regional elections, first in Lower Saxony and Hamburg and then nationally in the European elections. A loose alliance of green regional parties gained 3.2 per cent of the national vote and by 1980 the electoral potential was estimated to be 15 per cent. The Green Party of today was founded in order to participate in the General Election of October 1980.

Before this happened, however, some earlier right-wing influences were discarded. The party's platform remained strongly anti-nuclear and as such was totally opposed to the Bavarian Prime Minister, the CSU's Franz-Josef Strauss, becoming Federal Chancellor. The canditature for the Chancellorship became the main election issue in 1981, and was thus dominated by a battle of personalities. The SPD, led by Schmidt, won this contest and the Greens remained relatively unsuccessful with only 1.5 per cent of the vote. They were, however, more successful in those Länder dominated by nuclear politics.

In the unusual Federal elections of 1983, brought about by the fall of the Schmidt government, the Green Party only just managed to break the 5 per cent barrier and was able to send 27 members to Bonn. Although the acid-rain debate had by then reached a fever pitch, as described in Chapter 10, the threat of forest death did not lead to the feared electoral landslide for the Greens.

Table 5.1 shows the election results of major parties and the Greens between 1976 and 1987.

In the June 1984 European elections, the Greens gained 8.2 per cent of the national vote and the FDP disappeared from the European Parliament altogether. In 1985 the first red-green (SPD/Green) Land government was set up in Hessen, with the Greens in charge of energy and environmental policy, but excluding nuclear matters. This coalition collapsed in early 1987 after the Alkem scandal over the fate of reprocessed plutonium.

In 1986, the Greens adopted a programme for the transformation of industrial society, urging the recommunalization of the energy supply industry and its transformation into a non-profit making public service dedicated to energy saving. However, during the latter half of the 1980s, as the Green Party has had to participate in public debates about policy, it has become increasingly aware of the fundamental conflict between the technological and legislative

Table 5.1 Federal election results 1976–87

	CDU/CSU	SPD	FDP	Greens
1976	48.6	42.6	7.9	—
1980	44.5	42.7	10.6	1.5
1983	48.8	38.2	7.0	5.6
1987	44.3	37.0	9.1	8.3

responses it had helped to stimulate, and its own utopia. It found compromise difficult, and experienced damaging splits between fundamentalist 'Fundis' and pragmatic 'Realos'.

In early 1989, a Red-Green coalition began ruling in Berlin, but the Greens failed to be elected in the coal-mining Saarland in early 1990. By 1990, the environment had become subsumed into the national effort to unify the two German states and the CDU/CSU once again raised the Red-Green spectre as a threat to German stability.

A stabilization of the Green vote at around 10 per cent has taken place, but with considerable regional fluctuations. The Green Party seems to have become a small but stable feature of the political landscape, perhaps declining in influence at the Federal level, but useful in bringing minority concerns to the attention of the public and in keeping industry on its toes.

The Green Party, energy and environment

The Greens had participated in the energy-environment debate of the 1970s with rather utopian notions about the immediate shutting down of nuclear reactors and continued reliance on German coal. This hardly amounted to an energy policy acceptable to the majority of citizens, and even less to the Federal Government. At the regional level, especially in North Rhine-Westphalia, green groups supported strong demands for sulphur dioxide reductions prior to 1982. These reached a peak with actions by the small pressure group 'Robin Wood' in 1983.[22]

When forest die-back made headlines, the Greens saw their fears about the impact of man on nature confirmed. Their inevitable response was to face the electorate in 1983 with an energy policy which was both anti-coal and anti-nuclear. This implied draconian methods for improving efficiency in both energy supply and use. They also promoted policies threatening one of the great loves of most Germans, the unfettered possession and use of fast cars.

By increasing angst about nuclear power, the German Greens had played on the public's fear of ill health, genetic malformation, shorter life expectancy and eventual nuclear catastrophe. Their opponents responded with a more immediate, and perhaps more credible threat, that of an environmental and economic catastrophe within a decade through the loss of German forests. This held the promise of making nuclear power less reviled while weakening the economic viability of German coal.

The public became indeed somewhat more favourably disposed towards the construction of new nuclear plant. In early 1982, two-thirds were in favour of further plants being built, compared to less than half in early 1979.[23] Anti-nuclear protests declined during the early 1980s, but revived strongly after the Chernobyl accident, leading, in 1989, to the abandonment of spent fuel reprocessing inside the Republic.

Those environmentalists who took a serious interest in energy policy were also looking closely at the regulatory framework for the energy sector. Basing their arguments on developments in California, they looked for means to restrain energy demand. They saw the utilities as expansionist, wealthy and politically influential bodies weakly regulated by regional governments which were often major shareholders.

They drew attention to the fact that the utilities were still regulated by 'Nazi' law, the Energiewirtschaftgesetz (Energy Sector Law) of 1935, that industry was subsidized by smaller consumers and that some utilities were deeply involved in the nuclear cycle, from uranium mining through to reprocessing. The large regional electricity utilities thus made excellent targets for green propaganda. For environmental regulation in general, a degree of common ground between the Greens and a Federal bureaucracy interested in introducing more competition to the electricity sector was emerging.

In early 1988, a CDU environment minister called for a change in the Energy Sector Law so that environmental protection and energy efficiency would become a legal obligation for the utilities. With the emergence of global warming as a political issue, the debate about energy saving and energy efficiency has returned wholeheartedly to German politics. Energy experts who had once advised the Greens, now advise government.

Conclusions

The environmental movements in Britain and Germany, however different in form and channels of influence, have drawn attention to many of modern industrial society's problems and have offered solutions ranging from the utopian to the pragmatic.

In Britain, 'deep' Greens did so as a small component of a much broader environmental movement over which they had little influence. In Germany, circumstances were such that green ideas became widely influential as part of a broad protest movement which presented a direct political challenge to existing parties. The energy-environment debate of the 1970s formed a key part of that challenge.

However, the German Greens failed to halt the progress of nuclear power. Their primary aims, to force major improvements in energy efficiency and the abandonment of nuclear power, have not been fulfilled. The Green Party acted as an outsider addressing the public rather than as participants in the policy process. However, on acid rain, as described in later chapters, active environmental policies were advanced by both the Federal Government and politicians from the traditional parties. Ironically, green politics in Germany seems to have strengthened the state rather than weakened it.

The radicalization of German environmentalism can be ascribed, in part, to its lack of access to established channels of influence on public policy. It has few institutional links with the past and has lacked the desire and capacity to cooperate with the 'State'. Once the Green Party had become involved in party politics after the mid-1970s, German politics became more flexible, open, and unpredictable. The FR Germany became a more difficult partner in NATO and the European Community.

The energy-environment debate in Britain, on the other hand, did not become a major dimension of party-politics and the environmental movement could gain no advantages from supporting any one party. Policy was therefore bound to proceed more cautiously and modestly from the environmentalist perspective.

Green politics in the UK remained weaker because better integrated environmental groups, used to being consulted by government departments with respect to specific problems, could gain little support for acid-rain abatement policies from either senior politicians or civil servants.

During the early 1980s, public debates in Britain were dominated by economic issues. To the extent that energy and environmental issues were debated at all, they were subsumed by deregulation and privatization as the main concerns. As will be described in Chapter 9, the absence of any articulated energy policy, or the will to formulate one, denied the environmental movement any levers for influencing government decisions.

Notes

1. M. Hajer and M. Schwarz (1989), Acid rain and cultural climate: what is the problem?, Paper to the congress *Decisions Concerning the Environment*, Arc-et-Senans.
2. E. Müller (1986), *Innenwelt der Umweltpolitik*, Westdeutscher Verlag, Opladen.
3. C. Spretnak and C. Fridtjof (1984), *Green Politics*, Paladin, London; R. Bahro (1982), *Socialism and Survival*, Heretics Books, London.
4. G. Langguth (1984), *The Green Factor in German Politics* (translated Richard Straus), Westview Press, London.
5. T. O'Riordan (1981), *Environmentalism*, Pion, London.
6. H. Newby (1990), *Ecology, Amenity and Society: Social Science and Environmental Change*, Paper given to Royal Geographical Society/ESRC Conference, London, February 1990.
7. K-G. Wey (1982), *Umweltpolitik in Deutschland*, Westdeutscher Verlag, Opladen.
8. W. Huelsberg (1988), *The German Greens* (translated Fagan, G), Verso.
9. D. Vogel (1986), *National Styles of Regulation*, Cornell University Press.
10. J. Porritt (1984), *Seeing Green*, Basil Blackwell, Oxford.
11. E. Goldsmith (1988), *The Great U-turn: Deindustrialising Society*.
12. The description of the British environmental movement is based largely on P.D. Lowe and J. Goyder (1982), *Environmental Groups in Politics*, Allen & Unwin, London; M.J.C. Barker (1984), *Directory for the Environment: Organisations in Britain and Ireland 1984–85*, Routledge & Kegan, London; various profiles in Environmental Data Services Ltd, *ENDS Report*; and personal contacts and observations.
13. W. Rüdig (1981), *Die Politisierung der Atomkraft: Kernenergieentwicklung 1981*, Diplomarbeit, Freie Universität Berlin, p. 156.
14. N. Dudley, *et al* (1985), *The Acid Rain Controversy*, Earth Resources Research, London.
15. House of Commons Environment Committee (1984), *Acid Rain*, Fourth Report, Session 1983–84, HC 446, HMSO, London.
16. C. Rose and M. Nevill (1985), *Tree Dieback Survey: Final Report*, Friends of the Earth, London.
17. Proceedings of the Joint IEEP/NSCA Conference (1986), *Acid Rain: The Political Challenge*, 19 September, London.
18. W. Huelsberg (1988), *The German Greens*, Verso, London.
19. J. Radkau (1980), *Aufstieg und Kriese der deutschen Atomwirtschaft*, Rowohlt.

20. E. Brand (1988), *Staatsgewalt*, Die Werkstatt, Goettingen.
21. E. Wiedemann (1988), *Die Angste der Deutschen: Ein Volk in Moll*, Ullstein, Berlin.
22. For details on the responses by the Green movement to acid rain, see Arbeitskreis Chemische Industrie/Katalyse Umweltgruppe Köln (1983), *Das Waldsterben: Ursachen, Folgen, Gegenmassnahmen*, Kölner Volksblattverlag, Cologne.
23. P. Wiedemann and H. Jungermann (March 1989), Energy and the public, *Arbeiten zur Risiko-KommuniKation Heft 6*, Kernforschungsanlage Jülich.

6 Politics, government and environmental policy-making

Introduction

This chapter is concerned with those aspects of the political systems and government in the UK and the FR Germany which have had a significant bearing on environmental policy-making in general and acid-rain policy in particular.

Policy itself is a very broad concept which covers not only explicitly defined national objectives, but also guidelines, programmes and implicitly conveyed messages about desired goals and official preferences. It refers not only to the actions and objectives of central government, but also to those of lower tiers of government, e.g. Germany's Länder and Britain's counties, as well as those of government agencies. Although only part of the picture, 'official' policy as articulated by central government is nevertheless of great importance in channelling discussions with interested parties and pressure groups within a country, with the EC and with other trading partners.

Policy itself usually has its origin in the programmes of political parties. Policy thus has a political as well as an administrative dimension. The German language does not readily distinguish between policy and politics, the same word, Politik, being used to describe both. Politicians, once in government, may have to struggle for the acceptance of their policies against administrative or wider opposition. There is therefore no sharp distinction between the political and administrative processes in public policy-making which tends to take place at the interface between the two systems. The mechanisms for formulating and implementing policy may be formally legalistic, as tends to be the case in the FR Germany, or allow more political and administrative discretion, as in the UK. Power and influence in the making of policy thus falls to different professional groups and types of institutions in the two countries.

The policy-making function of government is difficult to separate from that of policy execution or administration. There are fundamental differences between the perceived role of government in Britain and Germany. The term 'state' (Staat) tends to be used more commonly in German than the term government (Regierung). 'State' is a broader concept which reflects the German emphasis on legal rules linking state to society.[1] In Britain, the concept of the state is weaker and government is seen as the instrument for implementing societal objectives rather than the automatic source of national leadership.

Three aspects of the political and administrative systems are considered in this chapter:

1 the political system, its components and underlying rules and traditions;
2 the structure of government and its constitutional basis; and
3 the administrative machinery for preparing and implementing environmental policy.

Party structure and electoral systems

The FR Germany

The German Parliament consists of a lower house, the Bundestag, and an
upper house, the Bundesrat (Federal Council). Members of the Bundesrat are
appointed by the regional parliaments which operate in each of the Federal
Republic's ten (now fifteen) Länder. Bundesrat representation is related to
population size.

Until recently, four parties dominated German political life. These were the
Christian Democratic Union (CDU), the Bavarian Christian Social Union
(CSU), the Social Democrats (SPD), and the small Free Democratic Party
(FDP). At the national level, the CDU/CSU tries to operate as a single party.

The history of German politics has necessarily been less continuous than in
the UK. Consequently, it is harder to identify natural constituencies for the
German political parties. The CDU and the SPD are best characterized as
mass parties which must gain their support through electoral competition over
policies rather than being able to rely on the automatic support of groups such
as the business community or the trade unions.

It has been rare for any single party to gain enough seats in the Bundestag to
form a government by itself. The rise of the Greens has further complicated the
bargaining needed to form coalition governments. As well as the Federal
Government in Bonn, Land governments need to be formed every four years.
For this reason, elections are virtually a continuous process in the Federal
Republic, making citizens constantly aware of political issues while the parties
themselves are continuously engaged in competition for public support. The
environment became a major element of this competition in the early 1980s.

To enter either Federal or Land parliaments a party requires 5 per cent of
the total vote. Once this initial hurdle is overcome, however, certain rights
follow, such as the availability of sufficient official funds to make participation in
political life relatively easy. This helped the Greens and increased the salience
of both energy and environmental affairs.

Party relationships are largely defined by the complex electoral system. One
half of the MPs are elected by direct majority on the basis of geographically
defined constituencies, while the others are elected by proportional representation on a common party ticket. Small parties generally enter the Bundestag on
the latter basis. The potential for instability inherent in coalition politics
significantly raises the importance of public opinion, large minorities and
electoral politics.[2] In general terms, few areas of German policy-making do not
become party-political issues.

This system encourages political flexibility and opportunism, but also consensus policies based on bargaining between coalition partners. It also encourages political responsibility on the part of those small parties which offer themselves as coalition partners. However, consensus politics during the 1960s and 1970s made it difficult for dissenting minorities to find modes of political expression. Extra-parliamentary protest politics developed after the end of the war, encompassing opposition to remilitarization during the 1950s, the student revolts of the late 1960s, the terrorism of the 1970s and the Citizens' Action groups of the 1980s.

There are slightly different legal and administrative systems in each of the Länder, as well as complex arrangements to prevent conflict between Federal and Land Law. Regional government is very strong and adds an additional layer of law, politics and bargaining to the policy formation process. As a layer of political activity which is closer to the grass roots, regional institutions help small parties to gain experience and base their appeal on more local isues. This factor too has encouraged environmental politics.

To counter the numerous divisions within German society, based on regionalism (almost tribalism), history, religion and class, great efforts were made in 1949 to prevent the re-emergence of either an unrestrained multi-party system or a strong central state. In very general terms, therefore, the German state is characterized by more checks and balances than its British equivalent. The dynamics of coalition politics and the federal system pervade the German acid-rain story.

The Union parties

Along with its coalition partner, the FDP, the CDU/CSU has been in power in Bonn since October 1982. The fact that they formed the government just as Waldsterben emerged as a major political issue symbolizes the acutely party-political dimension of acid-rain policy in the FR Germany.

The Union parties are united primarily by their opposition to socialism and communism but are not, unlike the Conservative Party in Britain during the 1980s, wedded to economic liberalism. However, anti-communism proved to be an increasingly meagre source of social cohesion and national commitment. In search of historical roots, the CDU/CSU looks back to the German Reich as created by Bismarck. While it is possible, from outside Germany, to consider the CDU/CSU as a single party, they are organizationally separate. Occasionally, the CSU threatens to set up branches outside its native Bavaria in opposition to its larger sister party.

The CSU/CDU's hankering for the Bismarckian solutions of the past (not shared by its coalition partner, the FDP) and its focus on the dangers of the formerly communist regimes to the east, were profoundly challenged by the Greens. The intense dislike of the German Union parties for the Greens must be emphasized as it accounts for some of the responsiveness of the established parties to environmental issues. In their early years, the Greens deeply offended many conservative politicians and their supporters.

With the passing of the Cold War, the Union has had to search for new unifying themes. Environmentalism has proved attractive given pressures against the open espousal of German nationalism. The Union parties' interest in the environment began in Bavaria, and reached a peak in the CSU's concern over acid rain. Until then, environmental protection had sat uneasily with the CDU/CSU's primary interest in stimulating economic growth.

The CDU/CSU were out of power from 1969 to 1982. By 1982, they needed to be responsive to public opinion but were burdened by their traditional support for nuclear power. Neither party had, in the past, shown a particularly strong interest in the environment. The CSU had an additional set of concerns about the subsidy by Bavarian electricity consumers of expensive German coal. This issue is discussed in Chapter 8.

The Social Democrats

The SPD has its roots in industrial mining areas. If the CDU looks back nostalgically to Bismarck's Reich, then the SPD looks back to the Weimar Republic. It was deeply disappointed about its failure to govern immediately after 1945 for, as a major opponent of the National Socialists, it had expected this almost as a right.[3] The SPD formally discarded its Marxist philosophy in the early 1960s and adopted the 'social market' doctrine, rejecting state ownership but remaining committed to intervention and regulation. Like the other mass parties, it had some difficulty in attracting and integrating the post-war generation into its ranks.

The SPD formed a 'grand coalition' with the CDU between 1966 and 1969 and succeeded in forming a government in its own right, though in coalition with the much smaller FDP, in 1969. At that point, it inherited a strong economy based on private industry and informal but important state-industry links which were protected by the FDP. The SPD/FDP coalition collapsed in October 1982 when the FDP withdrew its support, believing that the SPD would not support harsh economic measures which it believed necessary.

Legislative initiatives had become difficult for the SPD/FDP coalition because of a majority of conservative votes in the Bundesrat after 1976. By the end of the 1970s, the SPD had largely lost its image as a reforming social force. Weakened in 1974 by the resignation of Chancellor Brandt, known for his support for environmental issues, the SPD became associated by many with the virtual halt in environmental progress during the latter half of the 1970s.

The SPD developed a new environmental programme for the 1981 election, which did not, however, make any precise commitment about acid-rain abatement. The party now supports a gradual phase-out of nuclear power in the Federal Republic. SPD/Green coalitions have been formed in a number of Länder, starting with Hessen in 1985, Berlin in 1989 and, most recently and most significantly, in Schleswig-Holstein. Such alliances have kept the whole political system sensitive to green issues.

The Free Democratic Party (Liberals)

The history of German liberalism has been one of steady decline. From being the major parliamentary force in the 1870s, it was virtually eclipsed by 1932.

Since 1948 the FDP has broadly adopted the role of 'ideological flywheel' on the right with respect to economic policies and 'reasonable reformer' on the left with respect to social issues. It was formed in 1948 from the remnants of several Liberal parties with roots in middle-class Protestantism.

The FDP stresses its role as an integrative political force allowing the disenchanted (and generally well-educated) voter the means of protesting against the centralization of power. Its main problem is the maintenance of political support among members of the new middle class, for whose vote it must now compete with the Greens. For this reason, the FDP was particularly sensitive to environmental issues in 1981–2.

Criticized for opportunism, internal factionalism and disproportionate influence on German politics, its loyal supporters are rarely more than 5 per cent of the electorate and every election tends to be a struggle for survival.[4]

The FDP's political influence has been very large, however. As junior coalition partner, it has been in office since 1969. During its coalition with the SPD (1969–82), it consistently provided 4 out of a total 17 cabinet posts, while holding between 7 per cent and 10 per cent of the seats in the Bundestag. All the ministries most directly involved in the air-pollution control debate of the early 1980s were in FDP hands – Foreign Affairs, Economics, Agriculture and, most importantly, the Interior. All three Interior ministers between 1969 and 1982 were FDP members.

The impact of green politics on the parties

Until the mid-1970s environmental policy-making had been essentially an administrative task supported by enthusiastic politicians from the governing parties, such as Hans-Dietrich Genscher and Willy Brandt. As described in Chapter 9, ambitious legislative efforts were made. The differences between the early environmental programmes of each of the parties were marginal, resulting in a limited degree of political competition over green issues.

This changed as political instability increased during the late 1970s, with disputes over nuclear power and Ostpolitik. The links between environmentalism, green politics and the politics of protest strengthened. As the SPD/FDP coalition faced increasing economic difficulties, a stagnation in the environmental policy push, described as an 'ice age' by one observer,[5] set in. Large gaps in the implementation of existing environmental legislation became apparent and environmental issues became increasingly attractive to the Opposition.

By 1979 it had become clear to all of the established parties that their initial dismissive response to the Greens had failed. Academics began to suggest that a crisis of legitimacy faced the established political system. Political responses which would be seen as unifying and protective of the public interest were urgently needed. With the Green electoral challenge, an era of party-political response to the environment began in 1979–80. The small FDP successfully positioned itself as the most environmentally aware of the traditional parties and came to play a major role in the early acid-rain debate, a role subsequently reinforced by the CSU.

The British system

The fact that the British Parliament is elected under a winner-takes-all system results in great differences with the FR Germany. The consequent two-party system makes political expression for minorities or secondary issues such as the environment very difficult. Unlike in Germany, the British government is imbued with a considerable degree of stability irrespective of day-to-day public opinion.[6] Also, unlike in Germany, Parliament's upper house, the House of Lords, is not elected, represents no particular interest, regional or sectional, and has powers limited to delaying or revising legislation.

The fact that the government may call an election whenever it feels that it has the best chance of renewing its mandate (within the constraint that elections must be held at least once every five years), reinforces the ability of the government to define political priorities, which reflect the views of a single party.

While regional interests play a minor role in public policy-making, sectional interests are important. Examples include the protection which trade unions have, in the past, achieved for their members through the Labour Party and the influence of agricultural interests within the Conservative Party. Problems such as industrial pollution and the global environment attract political attention only with great difficulty. The prolonged political stability between national elections tends to lead to a weakening of party competition over policy for long stretches of time.

The two main parties, the Conservative and Labour Parties, bear some ideological similarities to their respective German counterparts, the CDU and the SPD. However, there are importance differences. The CDU sits with the centre-right grouping in the European Parliament while the Conservative Party is allied with the more unambiguously right-wing parties. In the centre ground of British politics lie the Liberal Democrats which do, however, play a considerably less important political role than does the FDP.

In spite of some success in the 1989 European elections, the Green Party is not a significant political force. However, its limited success, coupled with the Prime Minister's expressed interest in environmental affairs, has significantly raised the importance of the environmental issue in the UK and has ensured that competition between the parties is beginning to emerge.

The Conservative Party

The Conservative Party, which has been in power since 1979, has determined environmental, and hence acid-rain policy throughout the 1980s. Opposition disunity ensured that there was no plausible political challenge to Conservative rule. Environmental issues have tended to draw out the distinctions between the 'wet' and 'dry' (Thatcherite) wings of the party.

The Conservatives' traditional constituency is to be found among the landed aristocracy, the business community and the affluent middle classes. However, the party's mass electoral appeal lies in the expanding middle-class and the growing number of affluent working-class voters.

During the 1980s, the development of the Conservatives' environmental policy was subservient to other policy themes which included a fluctuating commitment to monetarist economic management, controlling inflation, lowering taxes, reducing the Public Sector Borrowing Requirement, reducing regulatory burdens and rolling back the state through the privatization of nationalized industries. The Conservatives also adopted a strongly pro-nuclear energy policy when they came to power in 1979, abandoning this only in 1989 during the process of electricity privatization. The degree to which environmentally ambitious policies could be achieved was primarily dependent on compatibility with these other more pressing objectives. Chapter 11 shows how other political concerns led to a negative policy stance on acid-rain abatement.

There was indeed a great deal of hostility between the 'dry' wing of the Conservative Party and the green movement. Nicholas Ridley, while Environment Secretary, offended environmentalists with his views that 'people are more important than concepts, particularly when these are intellectually flawed and based on hysteria rather than logic'.

The Thatcher revolution notwithstanding, there are strong environmental threads running through Conservative Party ideology, which are associated particularly with its longer rural and aristocratic tradition. Prior to the 1987 election, the junior Environment Minister, William Waldegrave, described the Conservatives as the party of conservation. Within the Department of the Environment (DoE), this minister consistently promoted more assertive environmental policies and began, as described in Chapter 5, to establish links with the environmental movement.

In 1984, the Bow Group, representing the unfashionable 'wet' wing of the party, recommended the introduction of strict controls on power stations.[7] As early as 1985, the Conservative think tank, the Centre for Policy Studies, argued that the party should 'expropriate the green issue' and that the lack of action on acid rain was a 'lamentable failure of imagination, as well, possibly of judgement'.[8]

The party's 1987 election manifesto placed great emphasis on the countryside and the national heritage, appealing to its natural constituency in the rural parts of England, but was cautious with regard to larger scale environmental problems, placing emphasis on the need for scientific justification for environmental controls.

A major change in policy presentation was signalled by the Prime Minister's well-publicized speech to the Royal Society in September 1988 in which she focused on the potential dangers of global warming and cited the environmental impacts of acid rain. The replacement of the somewhat abrasive Nicholas Ridley as Environment Secretary by the more persuasive Chris Patten in 1989 underlined the Conservative Party's desire to present its environmental policies to the electorate in a more attractive light.

However, there has been an acknowledgement within both wings of the Conservative Party that the environment is a political issue. The 'dry' wing of the party has begun to reconcile its ideology with environmental concern by focusing on the possible role of 'market instruments', such as pollution taxes and tradable emission permits, in regulating environmental impacts. In 1989, the Environment Secretary, Nicholas Ridley, published his personal views about the role of environmental issues and market-based controls in the context of longer Conservative traditions.[9]

The Labour Party

The Labour Party's policies for the environment while in office have differed little from those of the Conservative Party, reflecting the lack, until recently, of party political competition over green issues. The establishment of the DoE in 1970 was planned by the Labour Party towards the end of its 1964–70 spell in government. Given the country's severe economic plight, little priority was attached to the environment by the 1974–9 Labour government. It failed to respond to the Royal Commission on Environmental Pollution's 1976 report on air-pollution control which recommended a complete overhaul of the system. A reply to the report from the Conservative government in 1982 rejected most of its recommendations.

In the early 1980s, the Labour Party was beset by problems about its relationships with the trade unions which provide a significant proportion of its finance. It was considerably weakened by the defection of senior figures from the right-wing of the party to a small Social Democratic Party (SDP) in 1980. The Labour Party's position as a mass party was threatened by the perception that its support was shrinking to disadvantaged and disgruntled minorities, such as the unemployed, public-sector employees and ethnic minorities.

In the latter part of the 1980s, the Labour Party's prospects of regaining office revived as it discarded previous policies such as unilateral nuclear disarmament and the commitment to state ownership of key industries. Trade-union influence has also been weakened. The fundamental policy changes echo those accepted by the German SPD in Bad Godesberg in 1959. By 1990, the Labour Party enjoyed a considerable lead over the Conservative Party in opinion polls.

Like the Conservatives, the Labour Party has begun to emphasize environ-mental issues. The 1987 election manifesto promised the setting up of a Ministry of Environmental Protection with a remit to 'take positive action to safeguard the quality and safety of life' and unspecified 'action to deal with acid rain'. The Party's recent policy review process has given fundamental consideration to new environmental policies and institutional changes. The Socialist Environment and Resources Association (SERA) is a small internal pressure group pressing for policy changes from within the Party.

The Labour Party's last term of office pre-dated the emergence of the major problems of acid rain and climate change and it is not easy to predict how the Party would respond to environmental challenges while in power. The waning influence of the National Union of Mineworkers (see Chapter 8) probably gives the Party a freer hand in dealing with energy/environment issues than would have been the case in the past. The Labour Party is, however, now committed to phasing out nuclear power in Britain.

The centre parties

The formation of the SDP in 1980 by disgruntled figures from the Labour Party and its subsequent electoral alliance with the much older Liberal Party promised at one point to change the shape of British politics. However, the failure of the two parties to coalesce into an effective political force has allowed two-party politics to re-emerge. One rump of the SDP joined with the Liberals while other supporters withdrew from the party merger. The SDP was subsequently dissolved.

The ambition of the Alliance (now the Liberal Democrats) had been to hold the balance of power in a 'hung' Parliament and subsequently bargain for the introduction of a system of proportional representation which would increase their representation. The popularity of the centre parties had slumped so much by 1989 that they were outpolled by the Green Party in the European Election. However, support for the centre recovered somewhat by the time of the 1990 local elections.

Although the 'third force' in British politics appears to have failed, it was the centre parties which argued most strongly for higher standards of environmental protection. Tom Burke, leader of the Green Alliance, has attributed the growing emphasis on green issues in Britain during the 1980s to the emergence of the SDP.[10] Thus, the centre parties appear to have played an important, if indirect, role during the 1980s by raising the level of party competition over environmental issues.

The structure of government

German Federalism, law and the constitution

The powers of the German Länder which together make up the Federation are clearly defined and enshrined in a written Constitution (Grundgesetz, or basic law). Infringements can be challenged in a Constitutional Court. The Federal government may pass legislation only if empowered to do so by the Constitution. While the Bundesrat may frustrate legislation, it can be outvoted by a qualified majority in the Bundestag after complex procedures have been exhausted. Overall, the Bundesrat is more powerful than the House of Lords and regional interests are more strongly represented in the Federal Republic.

Under the Constitution, the Federal Government is allowed to expand its responsibilities to further the uniform and equal treatment of its citizens. There has been an increasing trend for environmental regulations to fall within the competence of the Federal government as opposed to the individual Länder. This expansion is encouraged by intergovernmental treaties, including the Treaty of Rome and more specific pieces of EC legislation. The overall structure of the political and legislative system, with its Federal–Land relationship, is not dissimilar to the European–national relationship operating within the European Community.

There is a distinction between civil, criminal and administrative law, each of which has its own set of courts at both the Land and Federal level. Environmental law falls principally within the administrative category, although there is a new developing section of the criminal law dealing with the environment.

Two types of controls are allowed under administrative law. The first type is a Verordnung (regulation) which is directly binding on all persons and bodies within the Federal Republic. Regulations generally require the approval of the Bundesrat. The Bundesrat's consent must be obtained to satisfy the Federal principle when the interests of individual Länder may be affected by a Verordnung which cannot be challenged in the courts.

The second type of control is a Verwaltungsvorschrift (administrative directive) which is binding only on the bodies, operating at the Land level, required to interpret and implement it. The implementation of a Verwaltungsvorschrift may be challenged in the administrative courts by interested parties with standing. Since they are not binding on polluters, they allow more discretion at the Länder level. Given a choice, industry will often prefer Verordnungen which establish clearer and more uniform rules. Leap-frogging environmental demands at the Land level made under a Verwaltungsvorschrift were one reason why electric utilities argued for Federal legislation on acid emissions during the late 1970s. The Bundesrat's consent is also required for most administrative directives.

Federal institutions, in stark contrast to Britain, remain relatively weak in terms of management powers. Rather their competences cover international activities, strategic planning at the national level and a considerable degree of fiscal responsibility. Federal bodies are therefore attracted to national issues which arouse sufficient public concern to justify Federal initiatives. The environment proved to be one such issue, creating a source of political energy for reform which was entirely lacking in Britain.

Federalism also means that regional boundaries are perceived to be important. 'Internal transfrontier pollution' from one region to another, e.g. from coal-producing North Rhine-Westphalia to Bavaria could therefore become a political issue. This strong regional element means that extensive bargaining is required to resolve conflicting interests. Conversely, any such environmental problem will invite politicization along regional lines, encouraging, in turn, Federal initiatives.

The West German ministries and the chancellor

In the West German system, the Chancellor who is the leader of the strongest party in the Bundestag, decides the principal policy guidelines (Richtlinien) on the basis of the programme adopted by party conferences. There is a direct, but not compelling link between party policy and government policy. As in Britain, most legislation originates from the Federal ministries rather than from individual MPs.

Ministers enjoy greater autonomy than in Britain. They also tend to remain in the same post for far longer than in the UK, allowing them to build up considerable degrees of expertise with respect to their particular portfolios.

Within general guidelines, ministers may determine the policies for which they alone are responsible on the basis of the Resortprinzip (departmental principle). There is therefore less emphasis on collective responsibility (the Kollegialprinzip) within the German cabinet. Ministers are expected to be experts in their area of competence and able to fight effectively for their ministerial interests, if necessary in the public arena.

However, in the case of inter-ministerial conflicts, decisions will have to be taken in cabinet on the basis of a compromise negotiated with the Chancellor in the role as broker, and, if necessary, the final decision-maker. Acid abatement decisions were taken at Cabinet level.

Agreement in Cabinet does not guarantee the final acceptance of new legislation which must gain support from Parliament. Policy may also be influenced through Parliamentary committees, i.e. for environmental policy, the Interior Committee until 1986 and the Environment Committee thereafter. The Bundestag strongly urged immediate action against acid rain in the early 1980s.

The German Länder

The Länder are responsible for virtually all of the implementation and enforcement of Federal legislation. They do so on the basis of their own highly valued legal traditions and institutional arrangements, many of which date back to the Middle Ages. Länder may legislate as long as their competence has not been pre-empted by existing Federal law. Even the municipalities remain independent except in so far as their actions are not constrained by Federal and Land law. However, the pressures for national uniformity are strong and are encouraged by a common bureaucratic tradition, legal education and the ownership patterns of German industry.

There is therefore a strong, unifying bond between the civil servants working for the Länder and the Federal government which counteracts the considerable 'tribal' element in German thought, institutions and politics.[11] Another bond, especially for technical experts working for government, springs from common membership of expert bodies like the Verein Deutscher Ingenieure (Association of German Engineers-VDI).

Cabinet government in Britain

Britain, almost uniquely among modern states, has no written constitution. The interaction and balance of powers between Parliament, 'government' as embodied in the cabinet headed by the prime minister, and the Crown has evolved over a period of centuries. The relations between British institutions are governed by a mutual acceptance of precedent and by the existing body of law as passed by Parliament and approved by the Crown.

The method of conducting government affairs has affected the ways in which decisions on acid rain were taken, allowing decisions to be made by a small number of people. Legal issues, or indeed Parliament, have been comparatively little involved in British policy-making in this area.

Much British environmental legislation, including that covering air pollution, is notably unspecific, its main function being to define the broad parameters by which regulations are made and to allocate responsibilities for implementation. In contrast to Germany, institutional allocations of competence, status and resources are therefore often more important than legislation itself.

Enabling legislation of this kind often allocates responsibility to a government minister and thus initiates a process by which the government of the day awards itself a set of executive powers which it requires to implement policy. Responsibilities for implementing legislation may also be allocated to semi-autonomous bodies such as the Health and Safety Executive or HM Inspectorate of Pollution (HMIP), or to local authorities.

Most legislation in Britain is now initiated by the government of the day and will be drafted by the appropriate government ministry rather than by individual members of Parliament. It is at this drafting stage that the wishes and ideas of the bureaucratic machinery affect policy.

New legislation on air pollution is relatively infrequent. Prior to drafting legislation, the government will generally issue one or more consultation document (green paper). Responses are then taken into account before submitting a bill to Parliament. Particular attention will be paid to responses from bodies with a direct interest in the legislation, such as trade associations, other government departments, the nationalized industries and certain non-governmental organizations. Although there is no statutory obligation for government to consult environmental groups, they may have an influence on policy as legislation is being drafted.

Executive decisions

Under the British system, the executive branch of government wields consider-able power. It has been customary to describe the central doctrine underlying the exercise of this power as being that of 'collective cabinet responsibility'. In principle, the cabinet meets, chaired by the prime minister as a 'first among equals', to arrive at a consensus on the matters under consideration.

However, this is an ideal view of cabinet decision-making. In practice, the locus of decision-making has moved towards whatever level has been most appropriate.[12] Decisions which involve only a single department, and which do not raise fundamental issues of government policy, will be taken solely by the minister concerned. Many routine decisions may even be taken by a junior minister or a senior official without the minister being informed.

In principle, the business of cabinet consists of two main elements: 'questions which engage the collective responsibility of government' and 'questions on which there is an unresolved conflict of interest between Departments'. However, through simple lack of time, the first element of cabinet business has practically withered away. Much of the government's strategic thinking is conducted by middle-ranking civil servants while, according to some observers,[13] senior civil servants spend much of their time protecting and coaching their ministers who rarely stay in a particular job for very long.

The longer term decline in the role of the full cabinet means that power has moved towards the prime minister's Downing Street Office and towards a large number of cabinet committees comprising a handful of ministers. Even the existence of a cabinet committee is notionally an official secret, although they are now increasingly being identified. Many cabinet committee decisions are not referred to full cabinet.

Like the German chancellor, the prime minister can wield enormous power in setting the agenda for full cabinet meetings, establishing cabinet committees and selecting their membership.

An increasing feature, typical only of the British government and particularly prominent during the Wilson and Thatcher administrations, has been the tendency to take decisions in informally constituted groups of ministers. This can vary from an 'inner cabinet' taking the major policy decisions to groups of ministers, officials and persons from outside the formal government machine meeting away from Whitehall to take decisions in less critical policy areas. Britain's decisions on acid rain fall into this category and were, at different times, taken both by cabinet committee and in less formal settings.

Local government in Britain

There are two tiers of local government in Britain. These are the counties (or regions in Scotland) and the boroughs (towns) and districts (rural areas). The upper tier of local government in the major conurbations was abolished in 1985.

The upper tier is responsible for running the police, the health service and the education system. The remaining responsibilities belonging to borough and district councils include road repairs, sanitation, the granting of planning consents for industrial, commercial and domestic premises and, most relevant from the point of view of this book, the control of pollution from technically simple industrial processes.

The activities of local authorities in the UK are very much constrained by spending limits set by central government. Local government and the regions can therefore be largely ignored in British 'acid-rain' politics, very much in contrast to the Federal Republic.

The administration of the environment

The German ministries

The sixteen Federal ministries and the Office of the Chancellor have all had an interest in environmental policy since the early 1970s.

1969 was the year in which environmental protection emerged as a duty for the State. From that point onwards, pollution control responsibilities began to accummulate in the hands of the influential Bundesministerium des Innern (Federal Interior Ministry-BMI) which took over the Ministry of Health's duties for developing anticipatory measures relating to air pollution.

The responsibility for air, water, noise, radioactive materials, wastes and town planning then lay with the BMI until 1986. In that year, a significant re-alignment of environmental responsibilities took place. The Bundesministerium für Umwelt, Naturschutz und Reaktorsicherheit (Federal Ministry for Environment, Nature Protection and Reactor Safety-BMUNRS, also known as BMU) was created. It acquired the BMI's environmental duties and a responsibility for nature protection from the Bundesministerium für Landwirtschaft, Ernährung und Forsten (Federal Ministry for Agriculture, Nutrition and Forestry-BMELF). While the Bundesministerium für Forschung und Technologie (Federal Ministry for Research and Technology-BMFT) still promotes environmental technology, some nuclear protection functions were transferred to BMU.

Not all environmental responsibilities are in BMU's hands. BMELF retains control of pesticides and landscape issues together with the Ministry of Family, Youth and Health, which also has an interest in the regulation of hazardous chemicals.

The setting up of BMU post-dates the major policy debate over acid emission controls in the FR Germany. It was established by Chancellor Kohl after the Chernobyl accident and just ahead of a crucial election in Schleswig-Holstein. BMU is a small ministry but is growing rapidly. Whether the setting up of a separate environment ministry will ultimately result in a greater priority being attached to environmental issues is not yet clear. On the one hand, BMU has a clear mandate to press for higher standards of environmental protection. On the other hand, a small ministry, with less political weight than BMI, may have greater difficulty in making its voice heard and has an even greater need for allies.

The Federal Ministry for Economics (BMWi) has no direct environmental responsibility, but together with the Ministry of Finance (BMF), has an interest in the wider economic impacts of environmental controls.

There is no separate energy ministry in the FR Germany and BMWi is responsible for energy policy in addition to its wider brief. BMWi's role is somewhat different from that of the Department of Trade and Industry in the UK. While BMWi proved to be the main adversary of the BMI during the acid rain debate, some convergence of economic and environmental interests developed in 1982–3, allowing stricter air pollution controls to be introduced. The implications of German economic management for acid-rain controls are addressed in Chapter 7.

As in the UK, the Auswärtige Amt (Foreign Ministry-AA) also has an interest in environmental issues of an international or transboundary nature.

Environmental policy is coordinated through two committees, one acting at the political level, the other at the top level of the bureaucracy. The Cabinet Committee on Questions of the Environment is chaired by the chancellor and its prime duty is to resolve conflicts between ministries. Ministers are assisted by parliamentary secretaries of state who are political appointments. The second major co-ordination committee involves permanent departmental heads meeting irregularly under the chairmanship of the BMU's (formerly the BMI's) minister.

Given the need for coalition governments, the difficulty of policy co-ordination from the top and the fact that ministers tend to remain in their post for considerable lengths of time, the allocation of ministerial posts to particular parties is an important factor determining policy priorities. The allocation also gives a clue to the priorities of different coalition partners.

In 1982–3, when the SPD/FDP coalition was replaced by the CDU/ CSU/FDP coalition, a number of key ministerial positions changed hands. As a result of its declining influence, the FDP lost both the BMI and BMELF to the Bavarian-based CSU. The FDP retained control, however, of the BMWi and the AA. This ministerial line-up is critical in explaining the result of the forest-damage debate. Germany's principal perceived acid-pollution victims, the Bavarians, had control over the ministries responsible for forests and the environment after 1982. The small FDP, their role in German politics threatened by the Greens, controlled the economics ministry which ceased its opposition to acid-emission controls in 1983. Viewed against this background, the policies adopted by the FR Germany take on an air of inevitability.

Advice on environmental protection

German environmental policy planning takes place in consultation with the Länder environment ministries and the Umweltbundesamt (Federal Environmental Office-UBA) in Berlin. UBA, which was set up in 1974, is larger than the Environmental Ministry itself. Most of its employees are technical experts in a wide range of fields including science, engineering, law and economics. They help the BMU to draft legislation, follow developments in science, technology and politics, liaise with industry, inform the public and commission their own research.

Apart from UBA, the BMU also receives advice from a Rat der Sachverständigen für Umweltfragen-RSU (Advisory Council on Environmental Questions), an appointed body of established scientists from many disciplines, including the social sciences. In addition, there is input from special inter-parliamentary bodies, the Enquete Commissions, which are similar to Royal Commissions in Britain. These may call witnesses and publish reports making recommendations to government. A Commission on Energy and Environment, set up in 1978 to consider energy policy, was disbanded in 1983 because it could not reach any consensus on the future of nuclear power. It was re-established in 1986 to consider the global warming issue.

There are also permanent Bundestag committees, which ensure parliamentary participation in policy discussions at a technical level. The precise role played by all of these bodies with reference to air pollution and acid rain is examined in Chapter 9.

Decisions on acid rain, which were reached after several years of negotiations, especially with the Bundesrat, are described in more detail later. Senior members of the administration in BMI had prepared and promoted stringent air-pollution abatement measures since the early 1970s, long before the political system found the subject attractive and enabled a conclusion.

The Interior Ministry

Acid abatement policy in the FR Germany was formed under the aegis of BMI, prior to the establishment of BMU in 1986. Like the Environment secretary in the UK, the Interior minister had a wide range of responsibilities, being in charge of constitutional and labour law, internal security and the border police. The breadth of these duties provided many opportunities for trading off different policy pressures.

In particular, the Interior Ministry was also deeply involved in the nuclear debate through its regulation of radiation protection, nuclear safety and waste disposal. Because energy issues are not addressed in isolation in the FR Germany, environmental policy relating to the energy sector can therefore be considered to have been better 'integrated' for purely institutional reasons.

In 1970, the BMI's environmental responsibilities were still relatively slim. Its environment department was divided into two sections: Water; and Air and Noise. The air pollution section's responsibilities included legal matters and air-quality surveillance, entailing preparation of the technical guidelines for the abatement of air pollution.[14]

By October 1971 a section with responsibilities to plan policy initiatives and deal with general matters, such as principles, international relations and enforcement, was added. In 1972, a Reactor Safety and Radiation Protection section was set up and remained part of the Environment Department until 1976, when it achieved independent departmental status.

The status of the Air and Noise section was downgraded during the mid-1970s as the emphasis changed towards product-related air-pollution control measures. However, the division responsible for clean-air planning was given the task of planning an initiative on stationary sources and air quality.

Little change took place until 1984 when a new period of growth began for the Air and Noise section. In 1985, Air and Noise acquired an Immission protection and International affairs division as international negotiations began to dominate air-pollution control once again. A relatively flexible bureaucratic structure was able to respond quickly to new changes in policy emphasis.

The UK Department of the Environment

The DoE has primary responsibility for environmental policy in the UK, plus implementation responsibilities in England and Wales. The Scottish Office and the Department of the Environment for Ireland have comparable responsibilities in the other parts of the UK. The DoE as a whole is headed by a secretary of state, who is a member of the cabinet, supported by three ministers of state and four under secretaries with more specific responsibilities.

The details of British environmental policy are to be found in a series of 'Pollution Papers' published by the DoE. Several of these have referred directly or in passing to air-pollution control.

The DoE is not in any sense a department of 'Environmental Protection'. The 'environment' in the DoE's title covers the built, as well as the natural, environment. The Department was created in 1970 by combining the ministries of Housing and Local Government, Transport, and Public Building and Works. However, the Department of Transport (DTp) was pulled out and given a separate identity again in 1974. The largest task of the DoE is the central co-ordination of local authority affairs and a comparatively small proportion of its resources are devoted to protection of the natural environment.

While the DoE plays the primary role, responsibilities for certain aspects of environmental protection are allocated round Whitehall and in bodies with agency type powers. For example, the responsibility for controlling marine and coastal water pollution, pesticides and agricultural chemicals lies with the Ministry of Agriculture, Forests and Fisheries (MAFF), freshwater pollution is regulated by the National River Authority (NRA) and the Nuclear Installations Inspectorate (NII) regulates nuclear safety. The Forestry Commission reports to the Scottish Office.

As described in Chapter 9, routine responsibility for air-pollution control is in the hands of HMIP and the local authorities. These bodies must interpret vaguely defined principles of control such as 'best practicable means'. Since its formation in 1987, HMIP has been an integral division of the DoE. While reducing the burden on central resources, the dispersion of environmental responsibilities increases the difficulty of co-ordinating policy and setting priorities.

Two directorates within the DoE have responsibilities relating to policy for controlling atmospheric pollution, the Central Directorate on Environmental Protection (CDEP) and the Directorate for Air, Noise and Waste (DANW). CDEP comprises central functions such as the statistics and economics division, policy co-ordination and a Central Unit on Environment Issues (CUE) which has the role of liaison with other Whitehall departments.[15]

DANW is most directly concerned with air pollution. Until recently it had three divisions: Air Quality; Noise, Nuisance and Waste; and the Chief Scientist's Group. A fourth division, Global Atmosphere, has recently been added. The Air Quality division has been responsible for the acid-rain issue. It is further divided into a scientific unit, a domestic unit and an international unit which has taken responsibility for EC/UNECE negotiations. The scientific unit sponsors a considerable amount of external research and consultancy work rather than relying on internal capabilities.

Currently, the DoE's Chief Scientist also heads DANW. However, during much of the 1980s, the post of Chief Scientist was at the deputy secretary level. The post was down-graded to under secretary after the previous incumbent, Martin Holdgate, left to join the Geneva-based International Union for the Conservation of Nature (IUCN).

The DoE seeks technical advice from the Warren Spring Laboratory which also compiles the Department's environmental statistics. However, the Warren Spring Laboratory's work is primarily for the Department of Trade and Industry (DTI). In spite of some external help, the DoE's relatively small directorates are attempting to cover the same ground as both the relevant divisions of BMU in Germany and UBA which employs hundreds of professionals. A report issued through the DoE's Management Information System for Ministers (MINIS) in 1985 concluded that the department's resources are 'more or less adequate for routine administration of existing legislation, but are generally insufficient for new policy initiatives or to allow more than a reactive approach to new environmental problems and international developments'.[16] There is little to suggest that this position has changed.

The environment in Whitehall

Because environmental protection impinges on so many other fields of government concern, the DoE must consult and negotiate with a range of other government departments. The concerns of other departments have been and remain a major constraint on UK acid-rain policy. The most important departments are the Treasury, the Department of Energy (DEn), the DTp, the DTI, the Ministry for Agriculture, Food and Fisheries (MAFF) and the Foreign and Commonwealth Office (FCO). The role of the Prime Minister's Office in Downing Street is also significant. The Scottish Office, the Welsh Office and relevant departments of the Northern Ireland government would also expect to be consulted on policy.

The Treasury, the role of which is examined in Chapter 7, has a legitimate interest in the cost of environmental controls, particularly those incurred by public sector bodies such as the pre-privatization electricity industry or the car company British Leyland. It is the most powerful department of government and traditionally plays the role of sceptic when any proposal with public expenditure implications, whether related to the environment or any other field, is made.

The DEn has held responsibility for 'sponsoring' the nationalized energy industries but has more recently been concerned with guiding the gas and electricity industries into the private sector. Its role as a direct sponsor will diminish, but it may retain responsibilities in relation to the regulatory bodies set up to oversee the gas and electricity supply industries. There has been speculation that a Conservative government would abolish the DEn once the privatization programme is completed, returning its responsibilities to the DTI.

The DEn has been a powerful actor during the acid-rain debate, acting in defence of its interests where environmental controls were in potential conflict with its policy objectives for the energy sector. The depth of the DEn's knowledge of and its links to the energy sector have enabled it to take a more informed, and hence stronger position in inter-departmental discussions than has the DoE.

There have also been conflicts between the DoE and DTp, particularly on the politically contentious issue of road-building programmes. Unusually, given the doctrine of collective responsibility, the Environment Secretary attacked road traffic projections published by the DTp in 1989 as a justification for a major road-building programme as unrealistic. Until 1988, the DTp also strongly opposed EC proposals on vehicle emission standards which would have threatened lean-burn engine technology.

The DTI oversees industrial matters and has an interest in the costs of environmental controls and their impact on the competitiveness of British industry. Since deregulation has been a major policy theme in recent years, the DTI has not been a supporter of tighter environmental controls. Industrial interests lobbying the government would expect the DTI to provide a reasonably sympathetic ear. The DTI 'sponsored' the nationalized car industry up to the mid-1980s and would have been concerned about the effect of vehicle emission controls on its competitiveness. Latterly, the DTI has become more interested in the potential which markets for environmental protection equipment might offer to British companies.

The FCO is concerned about the effect of British air pollution on foreign relations. With EC and Scandinavian relations in mind, the FCO has leaned towards support for more ambitious environmental policies.

The potential for inter-departmental conflict over environmental policy is large. Attempts to integrate environmental concerns into other policy domains inevitably involve the DoE treading on other departments' 'turf'. Britain's failure to adopt pro-active environmental policies partly stems from this factor.

The role of advisory bodies

The Government has a number of other sources of political and scientific advice from Parliament, appointed commissions and statutory bodies.

A standing Royal Commission on Environmental Pollution (RCEP) was established in 1970 to 'advise on matters, both national and international, concerning the pollution of the environment; on the adequacy of research in this field; and the future possibilities of danger to the environment'.[17]

Appointees to Royal Commissions are the 'great and the good', people who have distinguished themselves in some way in their careers, or through public service, and who are perceived to be untainted by commercial or political influences. A Royal Commission has considerable powers – it can call people before it, call for evidence in writing and examine any documents or records which it sees fit.

Since 1970, the RCEP has touched on air pollution in several of its reports and its Fifth Report, published in 1976, was devoted exclusively to air pollution.[18] However, the government did not reply to the report until 1982, turning down most of its recommendations.[19] Partly because of the declining receptiveness of the government during the 1980s to advice from the 'great and the good', the influence of the RCEP has declined and later reports have had considerably less impact than those issued in the early to mid-1970s.

In 1978, to supplement the RCEP, the government established a Commission on Energy and the Environment (CENE). This published one report, on 'Coal and the Environment' in 1979,[20] before falling into abeyance.

Since the acid-rain issue became controversial, the DoE has adopted the practice of appointing independent review groups. Appointees to these are typically mid-career or senior scientists with a brief to provide advice of a purely scientific nature. Several review groups have studied different aspects of the acid-rain problem. More recently, a Climate Change Impacts Review Group has been set up.

Parliament has provided advice of a more policy oriented nature to the government. The House of Commons Environment Committee has a duty to 'examine the expenditure, administration and policy of the DoE and associated public bodies'. Like the DoE itself, the Committee's brief extends well beyond environmental protection. Since 1984, it has begun to emphasize the environment as opposed to other issues, partly because it has been easier to achieve common ground between members with a range of political affiliations. The Committee produced a stinging critique of government acid-rain policy in 1984 but subsequent reports have been less negative. Like other bodies, it has been able to build up its status by exploiting environmental issues neglected by the government.

The House of Lords European Communities Committee (Subcommittee F, formerly Subcommittee G) examines European environmental policy and reports on those which 'raise important questions of policy or principle.' There has been much overlap between membership of Subcommittee F and the RCEP. The House of Lords Committee has been critical of the Government with respect to both power plant and vehicle emissions but also, in the early 1980s, expressed considerable scepticism about European styles of regulation, notably the reliance on legal sanctions as opposed to the British traditions of persuasion.

One other important source of advice is the statutory Nature Conservancy Council (NCC). It has the task of providing national advice on the impact of pollution on habitats of protected fauna and flora. In 1984, while calling for more research, it argued that scientific observations supported the case for reducing acid emissions at source. However, this advice had little impact on the government. Under the Environment Protection Act, the NCC is to be broken up, with a separate Council being established for Scotland. This step has been critized by environmentalists.

Conclusions

While political and administrative structures served to encourage the early emergence of the environment as a major political and administrative issue in the FR Germany, similar developments have been greatly delayed in the UK. Several major asymmetries have been identified as being of relevance in accounting for the different priority attached to environmental policy in the two countries.

First, party political competition plays a greater role in West Germany. This makes the German system more sensitive to changing public moods. In turn, public opinion can become an instrument of government policy. Senior politicians in Germany were much more willing to communicate directly with the public about environmental issues and to encourage the perception of certain environmental threats. This willingness helped to engage public opinion and therefore promoted demands for environmental activism.[21]

In Britain, on the other hand, the dominant position of the single party in power lessens the need for interaction with the public. During the 1980s, government remained insulated from the development of a latent public concern about the environment and focused on a narrower range of economic and political issues.

The degree of centralization of political power also significantly affects the nature and quality of decision-making. In Britain, centralization is much greater, both geographically and institutionally, and decisions are based on narrower consultative procedures. Economic interests potentially affected by emission controls had more opportunity to express their objections in the UK and greater weight was attached to their views. The federal structure in Germany, on the other hand, tends to lead to complex, if slower decision-making processes involving a great deal of bargaining in the search for a wider consensus.[22]

The relatively important role of regional government in Germany has also encouraged the development of environmental policies. A great deal of routine work and implementation is handed 'down' to the regional level, allowing Bonn considerably more time to focus on strategic issues. Civil servants in London spend relatively more time serving the political needs of their ministers and attending to routine managerial matters.

Legislation and legal processes are more important in Germany. Law is widely perceived as being essential for both restraining and enabling political activity. The law defines not only the relationship between the Federal Government and the Länder, but also the rights and obligations of polluters and environmental administrators. In Britain, administrative bodies enjoy considerable powers of discretion in interpreting vaguely specified legal principles. Lack of assertion on the part of politically isolated administrative bodies discouraged the development of emission controls.

The Federal/Länder divide in Germany has encouraged competition between what has often amounted to dual legislative powers. The political energy thus generated provided further momentum for attempts to improve standards of environmental protection. This is explored further in Chapter 9.

A final difference between the countries is attributable to a traditional preference for generalists rather than specialists. Senior politicians and civil servants in Britain are regularly moved from one post to another, allowing them to accumulate very little basic knowledge of the work of their departments.[23] The permanency and size of the growing Federal bureaucracy dealing with environmental protection in Germany throughout the 1970s and 1980s has reinforced the wider significance of this policy realm.

Notes

1. J. Nevill (1982), *State and Government in the FRG: The Executive at Work*, Pergamon Press, 2nd edn.
2. E. Jesse (1986), *Die Demokratie der Bundesrepublik Deutschland*, Colloquium Verlag, Berlin (7th edn); and G.A. Craig (1983), *The Germans*, Meridian, New York.
3. A Ashenazi (1976), *Modern German Nationalism*, Schankmann.
4. G.K. Romoser and H.G. Wallace (1985), *West German Politics in the Mid-eighties*, Praeger, New York.
5. E. Müller (1986), *Die Innenwelt der Umweltpolitik*, Westdeutscher Verlag, Opladen.
6. Hirst has described this as 'the tendency of representative democracy to turn into "elective despotism"', see P. Hirst (1988), Representative democracy and its limits, *The Political Quarterly*, **59** (2), April–June 1988.
7. T. Paterson (1984), *A Role for Britain in the Acid Rainstorm*, Bow Publications Ltd, London.

8. Centre for Policy Studies, *Greening the Tories: New Policies on the Environment*, London, 1985.
9. N. Ridley *Policies against Pollution: The Conservative Record – and Principles*, Policy study no 107, Centre for Policy Studies, June 1989.
10. *ENDS Report 136* (May 1986), p. 12.
11. S. Greifenhausen and M. Greifenhausen (1981), *Ein Schwierig Vaterland. Zur politischen Kultur Deutschlands*, Fischer.
12. P. Hennessy (1986), *Cabinet*, Basil Blackwell, London.
13. Hennessy, op. cit.
14. This section is based on E. Müller (1986), *Die Innenwelt der Umweltpolitik*, Westdeutscher Verlag, Opladen. For a more historical dimension, see K.-G. Wey (1982), *Umweltpolitik in Deutschland*, Westdeutscher Verlag.
15. *ENDS Report* (1986), No. 135, April, p. 3.
16. *ENDS Report* (1985), No. 127, August, p. 3.
17. Royal Commission on Environmental Pollution (1971), *First Report*, Cmnd 4585 HMSO, London.
18. Royal Commission on Environmental Pollution (1976), *Fifth Report – Air Pollution Control: An Integrated Approach*, Cmnd 6371, HMSO, London.
19. Department of the Environment (1982), *Air Pollution Control*, Pollution Paper No. 18, HMSO, London.
20. Commission on Energy and the Environment (1981), *Coal and the Environment*, HMSO, London.
21. H. Weidner (1989), Die Umweltpolitik der konservativ-liberalen Regierung, and E. Müller (1989), Sozial-liberale Umweltpolitik, both in *Aus Politik und Zeitgeschichte, Beilage zu Das Parliament* 47–48/89, 17 November.
22. R. Mayntz and H. Scharpf (1975), *Policy-making in the German Federal Bureaucracy*.
23. Since the beginning of the acid rain debate, while Britain has had five Secretaries of State for the Environment (Jenkins, Howe, Heseltine, Ridley and Patten), Germany has really had only two (Zimmermann and Topfer). A third man, Wallmann, stayed for only a very brief period.

7 The economic dimension

Introduction

The cost of alleviating environmental problems such as acid rain runs into billions of pounds. Consequently, the degree to which measures to reduce atmospheric emissions were seen to be affordable has been a key question in the acid-rain debate, notably in Britain. The concept of affordability is far from simple. As argued in Chapter 2, policy decisions are seldom based on pure cost-benefit considerations. It is less a question of how much money is involved than who should pay, what flexibility they have, how they are constrained by their external economic and political relationships, how much environmental expenditure helps to meet other objectives, and what other expenditure options are foregone.

The central questions addressed in this chapter are: to what extent were Britain and Germany able and/or willing to bear the cost of cleaning up atmospheric emissions; and what aspects of the respective economic cultures facilitated or hindered the adoption of tighter emission controls?

In answering these questions, the emphasis in Britain must fall on the public sector, since most sulphur dioxide emissions, and a major proportion of nitrogen oxide emissions, have been the responsibility of industries in national ownership. In Germany, the issue is more complex. While it was mainly the private sector which was concerned, the industries involved tended to have a close relationship with the state which is, in any case, acutely concerned with directing the economy towards broader societal goals.

The following sections therefore discuss: the relative strengths of the German and British economies; the ownership and style of management associated with the industries contributing to acid rain; and the role of government in the economy, the relative importance of the different policy instruments available to it and the way in which it can influence the industrial sector.[1] The discussion focuses on generic differences between Britain and Germany. Detailed developments in sectors such as coal and electricity are described in Chapter 8.

Economic strength

The contrast between the economic performance of Britain and West Germany is stark. Rebuilding from an economy shattered by war in 1945, the growth of

Table 7.1   Measures of German and British economic performance in 1987

Measure	FR Germany	UK
Population	61.2 m	56.8 m
GDP at current exchange rates	$1117.78bn	$669.78bn
GDP at current purchasing power parity	$814.7bn	$702.5bn
GDP/capita (purchasing power basis)	$13,300	$12,400
GDP growth 1960–73	4.7%	3.1%
GDP growth 1973–82	1.8%	0.7%
GDP growth 1982–87	2.2%	3.0%
Average inflation 1973–87	3.7%	11.7%
Current inflation	3.1%	6.0%
Maximum inflation 1973–87	6.9%	28.1%
Unemployment as proportion of labour force	8.6%	11.6%
Investment in fixed capital as a proportion of GDP	19.3%	17.1%

Sources: Derived from Statistisches Jahrbuch, Umweltstatistischesamt; Annual Abstract of Statistics, HMSO.

gross domestic product (GDP) in West Germany has been almost 50 per cent more rapid than that in Britain. By 1987, GDP per head of population was some 50 per cent higher in dollar terms than in the UK when measured at current exchange rates. When account is taken of the relative purchasing powers of the deutschmark and the pound, Germans still appear to be some 8 per cent better off.[2] Table 7.1 shows a number of economic measures which readily distinguish British and German economic performance, while Table 7.2 shows the evolution of GDP per capita in both countries.

Environmental concern is recognized to be more marked among those with higher incomes for whom primary human needs are satisfied. To that extent, the greater degree of concern about environmental issues in West Germany which was identified in Chapter 4 may be linked partly to higher levels of income. German wealth may serve to strengthen a basic level of environmental awareness which exists for deeper cultural reasons.

Differences between the two economies are reflected in a variety of indicators other than wealth (Table 7.1)[3] German inflation averaged 4 per cent between 1973 and 1986, peaking at 7 per cent in 1974. In Britain, inflation averaged nearly 12 per cent over the same period, reaching 28 per cent in the mid 1970s and again passed the 10 per cent mark in 1990. The proportion of the workforce which was unemployed in 1987 was also lower in Germany (8.6 per cent) than it was in Britain (11.6 per cent), although both countries experienced huge rises in the number out of work from 1973 onwards.

The relative efficiencies of the workforces in Britain and Germany are sometimes offered as an explanation for superior German economic performance. German diligence is contrasted with the 'British disease', an inclination for tea-breaks and frequent strikes. This latter stereotype has been broken during the 1980s when GDP growth in Britain exceeded that in Germany.

Table 7.2   GDP per capita in Britain and Germany

	UK	FR Germany
1987	$12,400	$13,300
1980	$10,600	$12,000
1975	$9,700	$10,000
1970	$8,800	$9,200
1965	$8,000	$7,600
1960	$7,100	$6,300

Note: Based on price levels and purchasing power parity in 1987.

Sources: Derived from Statistisches Jahrbuch, Umweltstatistischesamt; Annual Abstract of Statistics, HMSO; OECD.

However, the most likely explanation for this latter development is that the British economy has had spare economic capacity after the very deep recession of 1980–81.

It is not necessary to resort to only cultural factors in order to explain Germany's relative economic success. The economic miracle was at least partially due to the injection of Marshall Aid and the arrival of large numbers of impoverished but well-educated and mobile refugees in the last 1940s. In addition, the capital stock in German industry was far less damaged at the end of the war than is commonly supposed. The post-war economic miracle provided the momentum for further rapid economic growth during the next three decades.

A further explanation is that Germany industry has concentrated on the production of capital goods, often with a high technological content. In the world economic boom of the early 1950s, this provided enormous potential for rapid growth. Certainly, Germany's better economic performance relative to Britain goes back to the late nineteenth century. Over-reliance on secure markets in the Empire and Commonwealth is sometimes offered as an explanation for Britain's poor comparative economic performance.

Other factors underlying the relative economic strengths of the two countries are perhaps more fundamental. The proportion of household income which is saved in Germany is much higher than in the UK, 15 per cent as opposed to 10 per cent in 1986. The German word for 'debt' (Schuld) comes from the same root as does the word for 'guilt'. Germans are far more reluctant to place themselves in a state of indebtedness than are the British, though this tendency is less marked among the young. The rate of saving in Germany is much more strongly influenced by the general level of economic activity than is the case in Britain. The British, on the other hand, tend to rely to a greater extent on institutional pension schemes to provide for the future and invest in property to a far greater extent than Germans.

The greater importance attached to the conscious act of foregoing consumption in order to provide for the future in Germany finds echoes in the 'Vorsorge' or precautionary principle applied to the protection of the environment from long-term damage.

The high savings rate is a source of economic vigour, as it is reflected in a higher rate of investment in new capital. During the 1960s and 1970s, as much as 25 per cent of German GDP has been invested in fixed capital, whereas, in Britain, the comparable proportion has typically been about 15 per cent. However, the current rates of investment have converged somewhat in recent years.

Thus, starting from a higher GDP base, the stock of industrial plant in West Germany is replenished more rapidly than in Britain, leading to a more modern capital stock which incorporates more productive, technologically advanced vintages of equipment. The government sees it as its task to ensure that this process continues as smoothly as possible.

As will be described, the political desire for environmental protection, entailing large capital expenditures, could more easily be accommodated in the German economic system which has a built-in tendency to adopt advanced industrial technologies and longer-term investment strategies.

Industrial organization

Hidden planning in Germany

Traditionally, German industry has been highly concentrated with a high degree of vertical integration in major firms. A relatively small number of large, highly diversified industrial concerns dominate many markets as well as accounting for significant proportions of employment. The top 50 enterprises in Germany account for approximately 50 per cent of total industrial turnover.[4] Attempts by the Allies at 'decartelization' in order to break up major blocs of economic power during the 1945–8 occupation had some success, but in many sectors, including coal, the longer-term trends later re-asserted themselves.

Investment is made with a longer term perspective in German industry than in Britain. This is attributable to habits which are deeply embedded in the patterns of finance and management of major German companies.

A difference between Britain and Germany is the relative degree to which industrial companies rely on equity and debt to finance their activities. British companies rely primarily on equity capital, while debt financing plays a far larger role in the Federal Republic. This points towards the great influence enjoyed by the 'big three' German banks, the Deutsche Bank, the Dresdner Bank and the Commerzbank. However, the banks' influence derives not so much from their loans, as from the fact they are also shareholders in German industry.

The total direct holdings are small, of the order of 10 per cent. However, the banks' influence is far greater than these holdings would suggest, thanks to the 'deposit voting right' (Depotstimmrecht). This allows banks to exercise a proxy voting right on behalf of customers who have made security deposits. The banks may also lend each other voting rights. It has been estimated that, through these devices, the private banks control 50–60 per cent of the voting rights exercisable at the annual general meetings of top companies.[5]

Two-thirds of these voting rights are concentrated in the big three banks. Since important company policy must be approved by at least 75 per cent of shareholders, major institutions often aim to gain at least 25 per cent of shareholders' voting rights in company in order to guarantee the power to veto certain decisions.

The banks' influence extends to more direct influence over the management process. Aktiengesellschaften (AGs), limited liability companies with publicly quoted shares, are the most common legal form for large German enterprises. These have a two-tier board structure, with separate executive and supervisory boards. The responsibilities of the supervisory board include the agreement of long-term capital expenditure plans and the appointment of the management board which has day-to-day responsibility for running the company.

Shareholders elect half of the seats on the supervisory boards of companies with more than 2000 employees, the other half being elected by the workforce. Banks account for a large proportion of the shareholders' representatives in the largest companies. Even more importantly, about half the chairmanships of the boards are held by bank representatives.

In taking an equity share in a company, banks are therefore doing much more than seeking a return on capital. They effectively participate in the running of the company and hence tend to take a long-term view of business prospects. Thus, the banking system had an intimate knowledge of the investment programmes of both the energy and the car industries and hence the potential impacts of environmental controls. The banks also bring to management another source of public influence, namely the views of small investors.

Those individuals who may be on the board of several major companies (bankers are permitted a maximum of ten directorships) constitute a highly influential network of financial power and advice, West Germany's so-called 'hidden planners'. This operates as an important stabilizing and co-ordinating mechanism within the West German business sector, since these bankers feel bound to use their influence to promote the harmonious functioning of the economy. If the idea that the promotion of a particular industrial sector may offer major opportunities takes root within this network of influence, considerable momentum for development is created. This has happened in the past with coal and iron and steel, and subsequently with nuclear power, information technology and, most recently and most significantly in the context of the acid-rain issue, environmental protection.

Short-termism in Britain

The links between the financial community and the industrial sector are more tenuous in Britain. The degree of participation experienced in West Germany is virtually unknown. The lower degree of concentration in the financial services sector and the importance of London as a financial centre is linked to a vigorous trade in equities on the Stock Exchange. Consequently, short-term market considerations tend to dominate in determining the availability of capital for major companies. Much of the British financial sector has had a trans-atlantic rather than a European orientation, though the Single European

Act is beginning to change this. To the extent that the financial community has thought about environmental protection at all, it has been more as an irritant which goes against the deregulatory trend which has dominated in the UK and the US during the 1980s.

Nationalized industries in the UK

In Britain, the determination of policy regarding acid rain did not impinge directly on private sector companies. With the coal and electricity industries in public ownership during the 1980s, policy-making was partially determined by the distinct framework within which the nationalized industries have operated.

The nationalized industries in Britain owe their existence to the Labour Party's traditional objective, expressed in Clause 4 of its constitution (now revised), to bring the commanding heights of the economy into collective ownership. Until the 1980s, the Conservative Party did not contest the public ownership of utilities and the coal industry, although the steel industry was nationalized and denationalized twice. The present government's privatization programme spells the end of the nationalized industry in the UK, at least in its traditional form.

The primary financial objective of the nationalized industries in Britain has been to break even 'taking one year with another'.[6] Many nationalized industries, notably British Coal, have conspicuously failed to reach even this vaguely defined objective. The nationalization statutes also require the industries concerned to take whatever steps are required of them by ministers in the national interest. Because of this, central government has become the ultimate manager of the nationalized industries, the programmes of which have often been subject to tight political constraints.

During the 1960s and 1970s, a more comprehensive framework of government financial controls emerged with three main elements: the financial target, the external financing limit and the performance aims.

The financial target is a medium-term instrument of control defining the rate of return on assets which an industry should achieve. The external financing limit is a short-term instrument which puts a ceiling on the amount which an industry may borrow in any given year. It is the device which the Treasury uses to help regulate the nationalized industries' contribution to the Public Sector Borrowing Requirement (PSBR).

The Treasury must also approve the capital expenditure plans of the nationalized industries,. Treasury approval was therefore needed for any spending on emission control equipment in the electricity supply industry during the 1980s. Major capital projects will get approval only if they are expected to earn a minimum rate of return on capital defined by the Treasury, known as the test discount rate. This has varied between 5 and 10 per cent in real terms (inflation adjusted) in recent years and is intended to reflect the opportunity cost of capital investment in the private sector. The current rate is 8 per cent. The nationalized industries have been able to pass the costs of unprofitable investments on to consumers, or have them borne by the taxpayer via the Treasury.

There is little doubt that parts of the nationalized industries are, in private sector terms, bankrupt. The burden of debts accumulated through unprofitable investments must be passed on to consumers unless the Treasury is willing to allow persistent losses or a debt write-off. This occurred recently in the case of British Coal. The nuclear power industry has also been unprofitable, though this may be partly attributable to political influence over choices of technology.

In principle, the government does not interfere in the day-to-day running of the nationalized industries. The reality is sometimes different. In the late 1970s, the Labour government, for example, persuaded British Gas to restrain rises in the price of household gas. In the early 1980s, the Conservative government was able to exert similar controls over industrial gas prices. External financing limits have proved an invaluable lever for government wishing to exert influence.

The government's ultimate power lies in making appointments to the boards of the nationalized industry. The choice of chairman is particularly important. Some chairmen can enjoy several consecutive terms of office. Others, who have resisted government pressure to implement particular policies, may enjoy shorter tenure.

The absence of a developed legal framework within which government's control of the nationalized industries might have been kept under judicial review has also helped to create a climate for government-nationalized industry relationships within which politicians have enjoyed very direct powers.

Ironically, a wide range of the political spectrum in West Germany, including sections of the CDU, was also in favour of nationalization in the late 1940s. However, plans were thwarted by the determination of the Allies not to allow the emergence of significant concentrations of economic power in post-war Germany.

The privatization programme

The very existence of the nationalized industries does not form part of the vision of Britain which has been pursued during the three Thatcher administrations. Based on economic ideas which were published by the Conservative think tanks in the 1970s, a comprehensive privatization programme has been implemented. By the early 1990s, the only remaining nationalized industries will be rail, coal and the Post Office.

The privatization of the state-owned sector of the British car industry was completed in 1988. The lengthy and complicated process of electricity privatization began in February 1988 and will culminate in the flotation of the two large generating companies in England and Wales in February 1991. The Conservative Party has promised to privatize the coal industry if it wins the next election.

The privatization programme has had four inter-related objectives: to roll back the state; to introduce competition into industries which have functioned as monopolies; to widen share ownership; and to raise cash which has helped to pay for tax cuts. These different objectives are often in tension with each other. Revenue from a privatization sale would often be maximized if competition were restricted. The sale of the gas monopoly in 1986, for example, was undertaken without any alterations to the industry's structure. Competition has subsequently been introduced by tightening up the regulatory regime. On the other hand, drastic changes are being made to the structure of the electricity industry which has turned out to be, along with water, one of the most controversial and problematical privatizations, principally because of the government's doomed attempt to reconcile competition in bulk electricity supply with support for the uneconomic nuclear industry.

During the 1980s, the *process* of preparing the nationalized industries for the market has been at least as important as the longer-term objective of enhancing competition. The preparatory process has involved restructuring the finances of the industries. Prices have been increased to improve the overall rate of return on capital and run down debt. In the case of electricity, negative external financing limits were set allowing all debt to be written off. Thus, ironically, given the government's ultimate intention of freeing the industries from state control, the 1980s saw even tighter controls on the nationalized industries, particularly on their finances. This became an important factor in determining the degree of acceptability of major environmental protection expenditure.

The government sector

'Government' has a twofold significance from the economic point of view. First, including regional and local government, it forms an important economic sector in its own right – accounting for almost 50 per cent of GDP in both Britain and West Germany. Second, 'government' has a more readily recognizable political role, which includes, among its many tasks, the management of the wider economy through fiscal, monetary and more direct types of intervention. The characteristics of 'government' as an economic sector (which are, in the German case, constrained by the Constitution) inevitably shape the economic instruments which are available to national governments.

A key characteristic distinguishing the public sectors in the two countries relates to the degree of political centralization. The concentration of economic power in central government is a strong feature of the British system where 80 per cent of tax revenue accrues to central government. The local authorities account for only 20 per cent of public expenditure, while 50 per cent of their income derives from grants from central government. The restriction of grants to local authorities according to spending targets defined by central government ('rate-capping' and subsequently 'poll-tax capping') has been an important instrument of central control in the 1980s. Thus, the fiscal authority of the British central government is virtually absolute.

In West Germany, on the other hand, the federal nature of the state is again of critical importance. In 1986, the Federal Goverment accounted for only 28 per cent of public expenditure, the remainder being accounted for chiefly by the Länder (27%), the municipalities (18%) and the separately administered social security system (38%).[7] Seventy-four per cent of the tax revenue in the Federal Republic is raised through joint taxes on income and sales which are shared by the different levels of government. The remaining taxes are levied directly by the Federal Government (12%), the Länder (5%) and the municipalities (9%). The level and distribution of the joint taxes are the subject of a two-year cycle of legislation requiring the consent of the Bundesrat. Thus, as intended by the authors of the constitution, federal control of fiscal policy in West Germany is weak, and an important economic policy instrument used extensively in Britain is essentially unavailable. The government must, therefore, use other means to exert influence over the national economy.

Economic management in the FR Germany

The significance of economic management for understanding the West German response to acid rain lies chiefly in the tradition of intervening in particular sectors, through regulation or financial assistance, either because of their strategic significance, or in order to secure broader economic objectives. This government agenda could often interact with that of the 'hidden planners' in the financial sector. The key point is that, in the German context, investment in pollution control equipment and associated technological developments could, in the correct circumstances, be viewed by government as a positive opportunity rather than as an economic handicap.

There has been greater continuity in the management of the West German economy over the last twenty years than in the UK, largely because of the more pluralistic environment within which economic policy is developed. In particular, the influence of the small FDP on economic policy cannot be underestimated. The FDP has been in coalition continuously since 1969 and, during that time, either the Finance or the Economics minister has been drawn from its ranks.

The guiding philosophy of all the major political parties is the 'Social Market Doctrine'. This mixes a belief in the desirability of markets and the pricing mechanism, concern for social justice and the compensation of the less well-off, and, where necessary, government guidance to secure control of the direction of the economy. Since the different elements of the 'Social Market Doctrine' clearly exist in tension with each other, there is ample scope for a diversity of political programmes which, however, remain true to the broad guiding principle.

More specifically, three distinct strands play a role in the complex set of interactions between the Federal Finance and Economics Ministries, the Bundesbank (which controls the money supply and credit), the Länder governments and the municipalities which, between them, determine economic policy.

The first element is the desire for a tight monetary policy and a dislike of deficit spending which can be traced to memories of the catastrophic hyper-inflation which foretold the end of the Weimar Republic. The 1949 Constitution provided no legal basis for deficit spending and, during the 1950s, under the direction of Economics Minister Erhard, large budget surpluses accumulated. Constitutional change in the late 1960s provided a legal basis for limited budget deficits and, for the first time in the Federal Republic, for the practice of the Keynsianism which had emerged as economic orthodoxy elsewhere in the market economies.

While sound money is still an important thread in economic policy, high budget deficits have emerged as a persistent feature of the West German system, partly because of the difficulty which a weak Federal Government has in either controlling the totality of public spending or implementing unpopular decisions. As with fiscal policy, the Federal government's control of monetary policy is tenuous. Primary responsibility lies with the Bundesbank which is required to support the general economic policies of the Federal government, but which is intended to be independent in its primary pursuits of safeguarding the deutschmark and regulating credit. Whereas the Treasury and the Bank of England tend to work in unison in the UK, there have, on occasions, been major policy rifts between the Bundesbank and the Federal Government in Germany.

Corporatism, i.e. co-operation and co-ordination between state and industry, is the second element of German economic management. This stems from a long German tradition of direct government participation in the economy. However, the experience of the Third Reich, and the unattractiveness of the former command and control economies on West German's eastern borders, means that participation is practised more subtly through the use of market-based instruments which leave final discretion with the private sector of the economy.

Partly dictated by the exigencies of post-war reconstruction, a tradition of intervention through tax allowances and investment incentives, such as grants and low-interest loans, directed at key sectors of the economy, has developed. In the 1950s, the targets were coal, iron and steel. In more recent times, there has been a fostering of investment in information and communications technology, nuclear power and, significantly from our point of view, pollution control. A left-over from the days of Marshall Aid, the European Recovery Plan fund, often provides the desired resources for injections into the environmental protection programme.

While the original focus was on removing bottlenecks in strategically important economic sectors, the later emphasis is on promoting new industries which will foster innovation and the efficiency of the economy as a whole, and which will allow West Germany to compete effectively in international markets. It is notable that there are again limits to the control which the Federal Government can exercise over this type of economic intervention, much of which is decided and implemented at the Land level.

The final strand of German economic policy is management of the business cycle. Until the 1960s, this was achieved entirely by monetary controls. The explicit adoption of Keynsian ideas (in the 1967 Stability and Growth law) was comparatively late and the instruments of control were peculiarly adapted to the West German situation. Given that the ability of the Federal Government to use fiscal measures to influence the macro-economy, as most other governments would do, is severely limited, business-cycle policy has been implemented through sector-oriented micro-economic measures.

Business-cycle policy thus tends to converge with the corporatist tradition, so that decisions concerning the broad stimulation of economic activity are closely linked to those concerning allocation of resources to particular sectors.

Economic management in Britain

In Britain, on the other hand, expensive pollution control measures, for either energy or vehicles, could not easily be accommodated in the economic agenda of the 1980s. The power to make British economic policy, as has already been suggested, is highly concentrated at the centre, particularly in the powerful Treasury.

The British counterpart of the German Federal Economics Ministry, the Department of Trade and Industry (DTI), lacks the political clout of its German opposite. The short-lived Department of Economic Affairs (DEA) created during the 1964–70 Wilson government was intended to counter-balance the power of the Treasury and undertake a role similar to that of the German Economic Ministry. However, ministerial infighting between the Treasury and the DEA led to its abolition by the incoming Conservative government in 1970.

The DTI continued to act as the 'sponsor' of many of the nationalized industries within the government system well into the 1980s. This task is carried out by the Department of Energy (DEn) on behalf of the nationalised energy industries. As such, one of the DTI's main tasks has been to petition the Treasury for funding to help ailing nationalized industries. Its attempts to protect the troubled government-owned car manufacturer, British Leyland, from, among other things, calls for the adoption of catalytic convertor technology led it to argue against tighter vehicle emission controls. Major industrial companies which felt that they might be adversely hit by stricter emission controls also made the DTI aware of their views.

With its role as supplicant on behalf of the nationalized industries receding, the DTI has begun to play a new role, reflected in a new logo (the 'enterprise' department) and a major internal re-organization. It has been trying to promote the environmental protection industry. An award scheme for the development of innovative pollution control processes has been initiated but, again, even if it wanted to, it could not employ the battery of more interventionist measures available to the German Economics Ministry without the unlikely permission of the Treasury.

The centralism inherent in the British system, coupled with the major political shift since 1979, has resulted in a significant change in the tone of economic policy, involving a move away from orthodox Keynsianism economic management towards a somewhat wavering commitment to monetarist policies. The explicit adoption of monetarist policies by the Thatcher administration in 1979 marked only the last stage in a longer process. The acceptance of economic policies imposed by the International Monetary Fund on the Labour Government in 1976 as the condition for a loan was an earlier indication of this long-term policy shift.

This institutional background, coupled with the close links which exist between the Treasury and the Bank of England, explains a bias in Britain towards the management of broad monetary and fiscal aggregates, as opposed to detailed sectoral intervention. The British economic management tradition discouraged the imposition of expensive environmental controls on the nationalized industries. On the other hand, German traditions served, in a positive way, to promote tighter environmental regulations.

Conclusions

Both the relative strength of the German economy during the 1980s, and the style of economic management reinforced wider pressures to improve standards of environmental protection.

While environmental controls may be an economic burden to those directly affected, they may also be an economic opportunity for certain industries and the wider economy. In Germany, the limited number of macro-economic instruments available to the Federal Government ensured that the stimulation of environmental investment through regulation was accepted, in the early 1980s, as an economic opportunity.

In Britain, economic difficulties and the susceptibility of the government to concerns about the profitability of key industries, many under national ownership, led to environmental controls being perceived primarily as an economic threat.

Notes

1. There is a very large literature on the British and German economies and their management. A short selection of recent English language material includes:

BRITAIN

D. Coates and J. Hillard (1987), *The Economic Revival of Modern Britain: The Debate between Left and Right*, Edward Elgar, Aldershot.
A. Gamble (1988), *The Free Economy and the Strong State: The Politics of Thatcherism*, Macmillan Education, Basingstoke.
A. Glyn and J. Harrison (1980), *The British Economic Disaster*, Pluto Press, London.
S. Pollard (1982), *The Wasting of the British Economy: British Economic Policy 1945 to the Present*, Crook Helm, London.
G. Thompson, V. Brown and R. Levacic (eds) (1987), *Managing the UK Economy: Current Controversies*, Polity Press, Cambridge.

GERMANY

S. Bulmer (ed.) (1989), *The Changing Agenda for West German Public Policy*, Dartmouth, Aldershot.
L. Erhard (1975), *Prosperity through Competition*, Greenwood, first published by Thomas and Hudson, 1958.
J.H. Knott (1981), *Managing the German Economy*, Lexington Books.
E. Owen Smith (1983), *The West German Economy*, Crook Helm, London.

2. OECD (1987) National Accounts Statistics, Paris.
3. The figures are derived from the standard annual statistical publications in the two countries: Central Statistical Office, *Annual Abstract of Statistics*, HMSO, London; and, for the FR Germany, Umweltstatistisches Aamt, *Statistisches Jahrbuch für die Bundesrepublik Deutschland*, Kohlhammer, Stuttgart/Mainz.
4. E. Owen Smith (1983), *The West German Economy*, Crook Helm, London.
5. Owen Smith (1983), op. cit.

6. J. Surrey (1986), Government and the nationalised energy industries, in A. Harrison and J. Gretton (eds), *Energy UK 1986: An Economic, Social and Policy Audit*, Policy Journals, Newbury.
7. Umweltstatistisches Amt (1988), *Statistisches Jahrbuch für die Bundesrepublik Deutschland*, Kohlhammer, Stuttgart/Mainz.

8 The energy sector in Britain and Germany

Introduction

The reaction of the main energy industries, notably those companies and organizations which formed part of the coal-electricity-nuclear complex, has been a critical factor determining the resolution of the acid-rain debate. In addition, energy has been, and remains, an important political issue in itself. Germany and Britain share an important common feature, namely the social and economic problems associated with the declining production of high-cost domestically mined coal. Britain and Germany have also, with varying degrees of success, pursued the development of nuclear power, against considerable public opposition, with the aim of reducing the use of domestic coal in power stations. Interactions between these particular issues and the energy and environment debate have been an important feature of political debate in both countries during the 1970s and 1980s.

Energy plays a unique role in underpinning social and economic activity within all economies. The use of oil as a transport fuel facilitates commerce and permits personal mobility to a degree which was unknown a generation ago. Just as importantly, the role of electricity as a clean, effective carrier of high-grade energy has reduced industrial costs, permitted the use of labour-saving appliances in the home and allowed the development of information and communication technologies which have significantly altered people's lifestyles. Thus, the use of energy is intimately associated with social and economic patterns.

At the same time, however, the production, conversion and distribution of energy has far-reaching environmental impacts. Many of today's most important environmental issues have their origins at least partly in the energy sector. As well as acid rain, the problems of global warming, the land impacts of coal mining, oil spillages, nuclear safety and radioactive waste disposal spring readily to mind.

Inevitable conflicts between energy and environmental policy objectives have had to be reconciled in order to produce politically acceptable solutions to the problems of acid rain in Britain and Germany. Accordingly, in this chapter, the structure and characteristics of the energy sectors in Britain and Germany are discussed in some detail.

The chapter has the following structure. First, there is an overview of energy supply and demand within Britain and Germany, followed by a discussion of the evolution of energy policy. The bulk of the chapter is devoted to a more detailed examination of each of the major energy industries – coal, oil and gas, and electricity. Here, particular emphasis is placed on organizational structure and relationships with government.[1]

Energy demand in Britain and Germany

Primary energy use in FR Germany was 430 million tonnes of coal equivalent (mtce) in 1987, compared to 343 mtce in Britain. Prior to the 1973 oil crisis, German energy demand grew at 4.7 per cent per year compared to only 1.6 per cent in the UK. The levels of energy demand in the two countries peaked in 1979 and, as a result of the deep recession of the early 1980s, remain lower than in 1973.

Over half the difference in final energy demand is accounted for by the industrial sector (Table 8.1). Industrial energy demand in Britain is now little over half the German level, reflecting mainly the declines in the energy-intensive industries, such as iron and steel, which took place in the early 1980s.

Patterns of energy use are most similar in the transport sector where, nevertheless, Germany still consumes 20 per cent more fuel than Britain. This primarily reflects a high level of car ownership. The use of energy in households and in commercial premises is 30 per cent higher in FR Germany, principally because of colder winters and higher comfort levels.

As striking as the differences in total energy use are the shares of each of the major fuels in the energy markets. The use of oil in Germany is almost twice as high as in the UK, accounting for more than half of final energy demand. German electricity use is almost half as high again as in Britain.

Table 8.1 Patterns of energy consumption in Britain and Germany in 1987 (in million tonnes of oil equivalent)

Demand sector	Britain	FR Germany	% Difference
Industry	33.5	59.6	+78
Transport	39.2	46.1	+18
Household/commercial	60.1	78.4	+30
Total energy	132.8	184.1	+39
Non-energy uses	12.5	16.3	+30
Statistical difference[a]	−0.5	−2.2	–
Total	144.8	198.2	+37
Transformation losses[b]	64.1	68.4	+7
Primary energy use	208.9	266.6	+28

Notes: (a) discrepancy between recorded demand and supply; (b) energy lost in power stations, petroleum refineries and secondary fuel plants.

Source: Eurostat Energy: Monthly Statistics, Office for Official Publications of the European Communities.

Table 8.2   The supply of energy in Britain and Germany in 1987
(million tonnes of oil equivalent)

Fuel	Britain	FR Germany	% Difference
Indigenous production			
Coal	59.5	75.1	+26
Oil	125.9	3.7	−97
Gas	39.3	12.8	−67
Electricity	15.3	34.7	+127
Total	240.0	126.3	−47
Trade[a]			
Coal	5.5	−0.9	−
Oil	−49.1	110.4	−
Gas	10.0	33.0	−
Electricity	1.0	0.3	−
Total	−32.6	144.6	−
Gross Inland Consumption	208.9	266.6	+28

Note: (a) Net imports. A negative number indicates net exports.

Source: Eurostat Energy: Monthly Statistics, Office for Official Publications of the European Communities.

Energy supply

While Germany was dependent on imports for more than half her energy needs in 1986, Table 8.2 shows that Britain was comfortably self-sufficient. British self-sufficiency, which has existed since 1979, is due to the production of North Sea oil which was first brought ashore in 1974. Fully 40 per cent of British oil production is exported, more than compensating for net imports of coal and gas.

Gas production in the British sector of the North Sea accounts for about 75 per cent of the country's needs. Britain's energy self-sufficiency is not a permanent feature, but may last until well into the 1990s depending on the extent of future exploration and production of oil and gas in the North Sea.

The only indigenous sources of energy in FR Germany are lignite, high cost bituminous coal and primary electricity, produced mainly from nuclear power.[2] There is also some gas production in Northern Germany which meets about 20 per cent of the country's needs. Of the half of West German energy consumption which is met by imports, 80 per cent is accounted for by oil, with most of the remainder consisting of natural gas. German concern about energy imports is more strategic than economic. A large trade surplus in manufactured goods compensates for the cost of energy imports. Britain, on the other hand, is habitually constrained by trade deficits.

Patterns of trade are crucial in explaining the structure of the energy industries in both countries. Almost all energy trade involves the hydrocarbon fuels, oil and gas. As far as coal and electricity are concerned, the two countries have been, at least up till now, effectively 'energy islands'.

The oil industry is dominated by the multinational companies over which national governments can exert influence through fiscal and regulatory policies but which they cannot, in the end, control. Natural gas, while found onshore and in the North Sea, is extracted primarily by the major oil companies. It is therefore the coal–electricity axis which has become the focus of government energy policies directed at supply, as opposed to demand, in both Britain and Germany. It is also the coal and electricity industries which are of key importance in devising policies to combat acid rain and other forms of atmospheric pollution.

Energy policy in Britain and Germany

Whether any country needs an energy policy which is separate and distinct from its general industrial policy is a matter of debate and to some extent depends on energy import dependence and general trade performance. However, the strategic importance of energy and the way in which it underpins most other economic activity has meant, particularly since the 1973 oil crisis, that most countries have articulated an explicit energy policy.

Broadly speaking, the principal objectives of any energy policy are to procure secure supplies and to do so at the lowest possible prices. The balance struck between security and cost depends very much on the changing circumstances of an individual country. Environmental objectives have become an increasingly important influence on energy policy formation.

In the UK, a separate Department of Energy (DEn) has responsibility for energy matters. In the FR Germany, energy policy is the responsibility of the Federal Economics Ministry (BMWi).

The evolution of energy policy

In the late 1940s, Europe was almost exclusively dependent on coal as an energy source. The only exception was the use of expensive oil as a transport fuel. The growing availability of cheap Middle East oil in the late 1950s and 1960s reduced European dependence on coal. During the 1960s, mines were shut and employment in the mining industry declined. Governments of the left and right presided over the run-down of coal output in both Britain and Germany. By the 1970s, the FR Germany and Britain were among the few European countries to retain coal industries of a significant size, with markets concentrated chiefly in the electricity sector and the iron and steel industry.

The 'oil shock' which followed the 1973 Arab–Israeli war engendered a climate of deep pessimism and made the world acutely aware of the degree to which it had become dependent on a commodity produced almost exclusively in a politically volatile region. Coinciding as it did with increasing public concern about the global environment prompted by the 1972 Stockholm Conference and the Club of Rome's 'Limits to Growth' report, the oil crisis came to be regarded as a manifestation of an acute physical scarcity of oil. Forecasts of world reserves of oil running out made a significant impact on the public consciousness. As noted in Chapter 4, Germans were more receptive to the 'Limits to Growth' message than was the British public, though this may also reflect the greater German dependence on energy imports.

After 1973, the tone of energy policy changed significantly. Renewable energy sources, along with conservation, were advocated by the environmental movement while, within the mainstream energy community, the increased use of nuclear energy and coal was promoted to substitute for oil. High projections of future energy demand, unrealistic in retrospect, apparently gave opportunities for the nuclear industry to expand, for the historic decline of coal use in countries such as Britain and Germany to be reversed, and still left space for energy conservation and renewables.

As the 1970s progressed, billions of dollars were poured into research and development projects designed to promote new technologies such as the fast breeder nuclear reactor and fusion. Huge investments were made in nuclear power plants and new coal mines.

However, the deep economic recession of the early 1980s caused energy demand to fall both because of reduced economic output and because a major economic restructuring changed the composition of output in favour of less energy hungry activities. Lower expectations of energy demand in the 1980s had important implications for the supply industries. In particular, while in the 1970s there was seen to be room for expansion in both the coal and the nuclear industries, under the new regime, these two energy sources came into direct competition with each other. In addition, nuclear power was burdened with a growing scepticism about safety and arrangements for nuclear waste disposal, partly as a legacy of the Three Mile Island accident in 1979. At the same time, environmental problems such as acid rain were a potential handicap to the competitiveness of the coal industry.

Thus, developments in the energy sector were leading to a conflict between supply sources into which environmental policy concerns, such as acid rain, would inevitably be drawn. As will be seen in Chapters 10 and 11, this issue became of considerable importance in Germany in the early 1980s while, in the UK, it was concern in the late 1980s about global warming which created very similar controversies.

Present energy policy

In both Britain and Germany, the role of market forces has been stressed by central government during the 1980s. In neither country does the government seek to impose rigid plans for the future on the energy industries. However, there are important differences in the way in which free-market rhetoric is interpreted.

In Britain, the DEn neither publishes energy forecasts nor refers to those made by outside bodies. The lack of any formal articulation of policy themes, even on an indicative basis, has been the subject of criticism from both the International Energy Agency and the European Commission.[3] The present laissez-faire presentation of energy policy is partly a reaction to the state planning flavour of policy during much of the 1970s.

In practice, the government cannot stand entirely aloof, as the continuing existence of a separate Department of Energy suggests. Decisions must regularly be taken on the oil licensing regime in the North Sea or on gas imports. In addition, some comparatively interventionist steps have been taken during the 1980s. These include the establishment of a grant scheme to promote coal firing in industry and the creation of an Energy Efficiency Office in 1983.

Much of the implicit energy policy during the 1980s has focused on the coal and electricity sectors. The promotion of nuclear power in the UK has been, until 1989, explicit and the measures taken to sustain its development have not, it might be argued, been in harmony with the government broader free market philosophy.

In the FR Germany, the guidance of the market in order to address broader social and economic goals through taxes, subsidies and voluntary agreements between industrial producers, negotiated at the behest of government, is well established. In this respect, the energy sector is no different from any other and the BMWi plays an important role in guiding its activities. Explicit energy programmes have been formulated which are underpinned by the usual devices of research and development support, soft loans and tax breaks.

Like the DEn in Britain, the BMWi does not publish its own energy forecasts. However, in articulating broad policy goals it does refer to forecasts made by outside bodies, including the German Economics Research Institute (Deutsches Institut für Wirtschaftsforschung-DIW).

Import dependence, the social consequences of the run-down of the coal industry, the development of nuclear power and environmental protection have all been elements of the energy policy articulated by the BMWi. FR Germany's 'hidden planners', described in Chapter 7, have played an important role in furthering many of the Federal government's energy policy objectives.

In the FR Germany, the Länder retain an important role in licensing new energy facilities and in regulating the price of electric power. Therefore, there is a strong regional flavour to energy policy which is absent in the UK.

A further development in the late 1980s has been the growing attempt to liberalize energy markets. The privatization of the British gas and electricity industries was one element of this wider push. At the European level, the objective of a single energy market by 1992 led to pressure on Member States to remove state aids and introduce competition. The German coal and electricity sectors may be significantly affected as a result of these changes.

Coal

Coal use has declined substantially in both Britain and Germany over the last thirty years. British coal output is now just over 100 m tonnes per year, little more than half the level in the early 1960s. Most of this decline took place during the 1960s while a Labour government was in office. It believed, like everyone else, that the period of low international oil prices would last indefinitely. The three major industrial disputes in the coal industry which have taken place in the last 20 years, all under Conservative administrations, have been major political events in the UK.

Table 8.3   Markets for coal in Britain and Germany 1988
(million tonnes of hard coal equivalent)

Market	Britain	FR Germany			% Difference
		Hard coal	Brown coal	Total	
Power Stations	82.5	47.0	28.0	75.0	−9
Other transformation	1.2	0.8	4.1	4.9	+308
Iron and steel	12.1	24.0	0.1	24.1	+99
Other industry	6.2	6.9	1.3	8.2	+32
Transport	–	0.1	–	0.1	–
Household/commercial	7.7	1.4	–	1.4	−82
Total	109.7	80.2	33.5[a]	113.7	+4

Note: (a) equivalent to 114.4 m tonnes of lignite mined in FR Germany.

Source: Eurostat Monthly Statistics.

Production of German hard coal has also declined, though not to the same extent. About 90 m tonnes per year is now produced, some 30–35 per cent less than in the early 1960s. The major producing areas are the Ruhr, part of the Land of North Rhine-Westphalia (NRW) and the Saarland, areas ruled by the SPD in the 1980s.

Fifty-five per cent of German hard coal is consumed in power stations compared to 80 per cent in the UK (Table 8.3). Use of coking coal in Britain is 50 per cent less than in Germany due to the much smaller size of the iron and steel industry. Coal consumption in industries other than iron and steel is broadly similar. The use of coal for space heating in households has virtually ceased in FR Germany while a small, but significant market remains in Britain.

Also of importance in Germany is the production of 'brown coal', or lignite; 115 m tonnes is produced annually in open-cast pits in the Ruhr and the area between Cologne and the Belgian border, also in NRW. Since lignite has a thermal content less than one-third that of hard coal, the 155 m tonnes of production is equivalent to only about 35 m tonnes of hard coal.

The German coal industry

Coal was the basis of traditional German industrial might. It made the Ruhr the heartland of German heavy industry. The power base of the SPD lies in this region. The coal industry now faces a major crisis. Production costs are high, markets are shrinking and long-term doubts exist over its largest remaining market, electricity supply.

Ruhrkohle AG, formed in 1969 by the amalgamation of a large number of mines, is responsible for 95 per cent of hard coal output. This amalgamation was forced by Federal law at a time when the mining industry was making considerable losses and there were no prospects of re-employment for unemployed miners. Ruhrkohle is owned by its major customers, the steel companies and the major utilities, VEBA and RWE among them. Job losses in mining are continuing, with 75 000 jobs having been shed since 1969.

The BMWi would like to see hard coal output cut drastically by 1995 when the main supply contract between coal and the electric utlities runs out. The fact that the European Community wishes to see state aids for the German coal industry removed increases the likelihood of this happening. However, there will be considerable opposition from the Land government in NRW. The matter has already reached the European Court, with Bonn supporting its coal industry.

Against this background, environmental controls which exacerbated the poor market position of the hard coal industry could be guaranteed to create controversy. Not surprisingly, those who in 1982–8 wanted to speed up the decline of the industry on economic grounds supported tighter environmental controls. In the coal-mining areas, a more complex debate took place. As will be described in Chapter 10, the mining unions at first opposed tight controls, but subsequently endorsed them on the grounds that measures which helped to give coal a clean image would enhance market prospects in the long-term.

FR Germany (even excluding the East) is the world's third largest producer of brown coal. In tonnage terms, ten times as much brown coal is mined in West Germany as open cast coal is mined in Britain. Entire villages are bought up and demolished in the path of the industry. Lignite production has been fairly stable over the last twenty-five years, though its use has declined by about 20 per cent in the last five years. As it is not economic to transport lignite any distance, most of it is therefore burned in power stations located close to the mines. The utility RWE owns 95 per cent of the total output through its wholly owned subsidiary, Rheinische Braunkohlenwerke AG. Brown coal is mined very cheaply.

British coal

The nationalized British Coal Corporation (BCC – formerly the National Coal Board or NCB), is responsible for most coal production in Britain. The NCB was created in 1948 from a wide range of private mining interests. Unsatisfactory safety performance and poor industrial relations in the private mines, as well as the broader desire to take control of the commanding heights of the economy, were the main political spur to nationalization. As in Germany the central problems over the last few decades have been high costs and the social problems associated with loss of employment. The number of miners fell from 583 000 in 1960 to 250 000 in 1974.

The first oil crisis was a turning point for the British coal industry. The 1973–4 miners' strike, which led to power cuts and three-day working in industry, precipitated the first general election of 1974. When the Labour Party narrowly won the election, the NCB and its employees had a unique opportunity to promote their industry.

In 1974, the coal Industry Tripartite Group was formed, comprising the government, the NCB and the coal unions. Against the background of a more general desire, at the national and international levels, to reduce oil dependency, the Group produced an ambitions 'Plan for Coal', foreseeing output recovering to 200 m tonnes per year by the end of the century.[4] This was to be achieved by: increasing coal burn in electricity supply; reversing the decline of coal use in industry; investing in new coal mines; and conducting R&D into advanced coal combustion technologies and the conversion of coal into gas and liquid fuels. At the time, oil and gas were believed to be so scarce that coal gasification and liquefaction would be needed by the mid-1990s.

In marketing terms, the Plan for Coal was unrealistic. However, partly as a consequence of higher oil prices, the decline in coal demand was arrested until the early 1980s. However, in 1980 the market for coking coal collapsed because of closures in the steel industry.

Against a background of high oil prices, the government introduced a Coal Firing Scheme in 1981 which provided generous grants for industrial companies wishing to convert to coal firing. However, in all other respects, the election of the new Conservative government in 1979 spelt nemesis for the British coal industry. Conservative leaders were acutely aware of their vulnerability to the power of the mining unions through their grip on fuel for electricity supply.

The plan for a massive programme of nuclear construction described below was part of a long-term effort to reduce coal dependency. At the same time, a shorter-term policy of building up coal stocks at power stations in order to ride out a national coal strike was instigated. The year-long 1984–5 miners' strike was widely regarded as inevitable given the antipathy between the Conservative government and the National Union of Mineworkers (NUM).

Partly as a result of the government's preparation and partly because of the extraordinary measures which the Central Electricity Generating Board (CEGB) took to maximize oil burn, the strike failed to produce power cuts, even when electricity demand reached a record high in December 1984.

The miners' strike was a watershed for the coal industry. The Nottinghamshire miners seceded from the NUM to form the Union of Democratic Mineworkers (UDM). With employees demoralized and divided, it became relatively easy to shut high-cost mines. Employment in the industry has fallen by 57 per cent since the end of the strike and the number of operating deep mines feel from 219 in 1979 to 94 in 1988.

Coupled with incentive schemes and a more towards flexible shift patterns, productivity has risen by 70 per cent since the end of the strike. In spite of the major drop in manpower, output is only 18 per cent below the 1983 level. The massive financial losses incurred during the 1970s and early 1980s have been significantly reduced to the point where the BCC has made an operating profit since financial year 1985/86.

However, it now appears that demand for coal could drop substantially during the 1990s because of the availability of natural gas for power generation, the possibility of imports of low-sulphur imported coal and the desire of the electricity supply industry to reduce its dependence on a dominant source of supply.

The oil and gas industries

The oil and gas industries in both Britain and Germany have been less central to the environmental policy debate than have coal and electricity. However, the environmental qualities and relative abundance of natural gas means that it is now a major competitor to coal in the power generation market. Natural gas is ash-free, sulphur-free, gives rise to lower nitrogen oxide emissions than either coal or oil and, similarly, entails lower carbon dioxide emissions.

Natural gas became available in the German market from finds in the Netherlands and Northern Germany in the 1960s. Gas also began to flow from the British sector of the North Sea in 1967. In both countries a rapid penetration of the household, commercial and industrial sectors took place. However, while almost all British homes have access to natural gas, this is not the case in Germany. Although Britain is not linked to the European gas grid, it receives over 20 per cent of its gas from the Norwegian sector of the North Sea. The FR Germany imports most of its gas from the Netherlands, Norway and the Soviet Union.

A substantial amount of gas was once burned in German power stations. While in Britain, the view was taken that priority should be given to supplying those sectors which could make the most economic use of gas's premium qualities, particularly the household sector.

After the 1973 oil crisis, policies in both Germany and Britain were aimed at minimizing gas use. A European Community Directive, agreed in 1976, restricted the use of gas in power stations. However, with the present abundance of natural gas, a reappraisal of its role in power station markets is now under way. The European Community has not rescinded the 1976 Directive, but has indicated that objections will not be raised if gas is burned in high-efficiency combined cycle power stations.

Oil

In both countries, the marketing and refining of oil is in the hands of the major multinational companies. The FR Germany has no major stake in the international oil industry and only 27 per cent of refining capacity is owned by German companies. Consequently the oil industry is foreign in German eyes. During the 1960s the Federal Government tried hard, though largely unsuccessfully, to create a German oil sector. The government's powers were constrained by European competition law.

After 1973, environmental controls were among the regulatory devices used in Germany to reduce the attractiveness of oil vis à vis alternative fuels, notably coal. For example, an Ordinance issued in the early 1970s required the use of flue gas desulphurization (FGD) at oil-fired power stations but not those fired on coal. New technical instructions on air-pollution control issued in 1989 restrict the use of residual fuel oil in combustion installations sized below 5 MW, while allowing the use of coal which is even dirtier, in terms of sulphur content.

There is a substantial British stake in the international oil industry through BP and the Royal Dutch Shell Group. The discovery of oil in the British sector of the North Sea in 1969 transformed not only the British energy sector but also the economy itself. Production started in 1975 and reached a peak a decade later. In 1984/85, over £12bn per year was being extracted from oil-related activity in the North Sea through royalties and taxes. However, the decision to price oil for the domestic market according to world prices means that the effect on the operation of final energy markets has not been pronounced.

In the UK, there are few restrictions on the quality of fuel oil which may be burned. While the European Community was able to issue a Directive on the sulphur content of gas oil in 1976, a parallel proposal covering the sulphur content of fuel oil was scrapped because of the lack of agreement between Member States.

Electricity

The great differences between the institutional arrangements for electricity supply in Britain and Germany are particularly important in explaining the different responses to environmental pressures. However, the basic economic parameters of the industries in terms of output, demand growth and fuelling patterns are also very different and have influenced the acid-rain debate. These aspects are described first.

Fuelling patterns for electricy supply

Whereas, back in 1960, electricity production in Britain was roughly 40 per cent higher than in FR Germany, the position has now reversed (Figures 8.1 and 8.2). German electricity demand has risen almost fourfold in the last twenty-five years, growing at 6 per cent per annum. About 150 mtce of energy, some 40 per cent of primary energy demand, is converted in power stations. In Britain, on the other hand, electricity demand growth has been little over 2 per cent per annum and just over 100 mtce of primary energy is used.

In 1960, hard coal or lignite accounted for more than 90 per cent of power station fuel in both countries. Germany's remaining electricity came from hydro-electric plants situated mainly in Bavaria, while Britain had already begun to make some use of cheap Middle-Eastern oil. Prior to the 1973 oil crisis, both countries attempted to make a greater use of cheap hydrocarbon fuels. In Germany, this primarily took the form of North Sea gas imported from the Netherlands. In Britain, fuel oil was used almost exclusively.

Central governments in both countries promoted the use of nuclear power. The British programme started earlier, and significant quantities of power were being produced by 1965. However, the British nuclear programme, reliant on two indigenous technologies, Magnox and Advanced Gas Cooled Reactors (AGRs), has not been a notable success and output was around 55 TWh in 1986.

However, in Germany, where use was made of adaptations of US-licensed systems, nuclear power production has grown to some 140 TWh per year. Nuclear power now accounts for 40 per cent of German electricity output (figure 8.1). This growth, which was strongly opposed by large sections of society, including the green movement, was particularly rapid during the early 1980s as the construction of Germany's convoy reactors (see below) got under way.

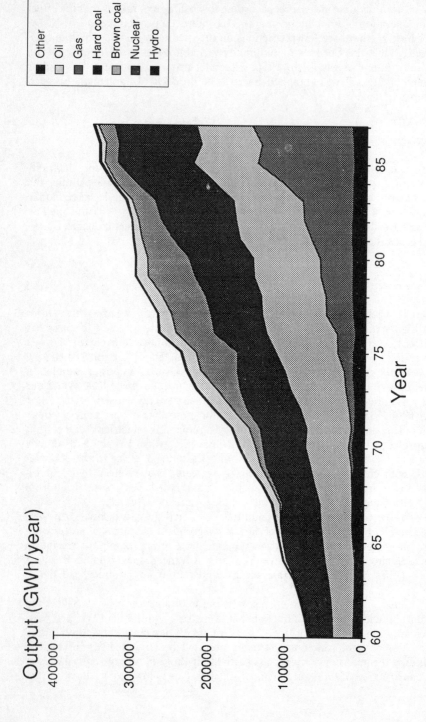

Figure 8.1 Fuel mix for electricity generation in the FR Germany

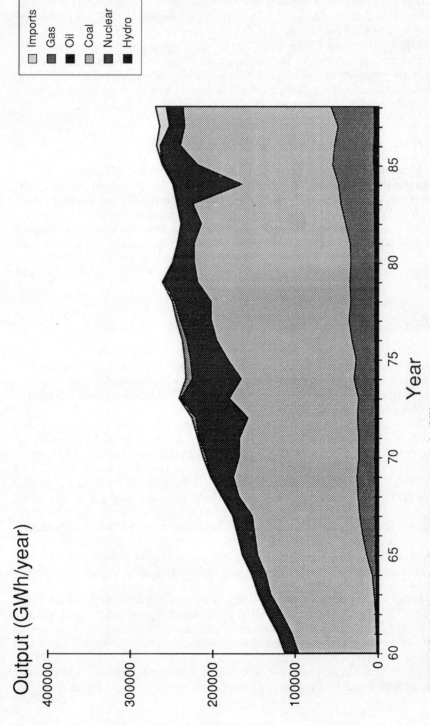

Figure 8.2 Fuel mix for electricity generation in the UK

The growth of nuclear power has weakened the importance of solid fuels in German electricity generation. The underlying competitive battle between coal and nuclear played a significant role in influencing the course of the acid-rain debate in Germany in the early 1980s. In Britain, on the other hand, the comparative failure of the nuclear programmes has left coal providing over 80 per cent of power station fuel requirements. The nuclear-coal issue did not directly influence the acid-rain debate in Britain but, as described in Chapter 11, was an element of the broader policy context.

Organization of electricy supply

The electricity supply industries (ESIs) in both Britain and Germany evolved from small, communally owned utilities created in the late nineteenth and early twentieth centuries. However, by the 1950s, entirely different structures had evolved. The British pattern was national ownership and control with responsibilities allocated to a relatively small number of bodies. However, in Germany a more fragmented system has persisted. A few vertically integrated utilities with well-defined territorial rights and obligations dominate the system. These are supplemented by hundreds of small, municipally owned utilities buying power from the majors and distributing it to final consumers.

Electricity supply in the UK

In March 1990, new electricity supply companies were vested in Britain in preparation for privatization. The new structure of the industry is discussed in Chapter 13. In this chapter, the pre-privatization structure of the industry, which is most relevant in understanding Britain's acid-rain policy during the 1980s, is described.

The move towards national ownership of the ESI began in 1926 with the creation of the Central Electricity Board which had responsibility for operating the newly created national grid. Full nationalization took place in 1947 and a further re-organization in 1957.

The most powerful actor within the nationalized ESI in England and Wales was the CEGB which generated most of the power consumed and operated the high-voltage super-grid. In addition, there were three vertically integrated utilities operating in Scotland and Northern Ireland – the South of Scotland Electricity Board (SSEB), the North of Scotland Hydro Electric Board (NSHEB) and Northern Ireland Electricity (NIE). The SSEB and NSHEB effectively operated as single system and had links with the CEGB: NIE forms an isolated power system.

Installed generating capacity in 1988 by company and plant type is shown in Table 8.4. This demonstrates the high dependence on coal which is particularly marked in England and Wales. Since England and Wales accounts for nearly 90 per cent of power generation, the central focus in this section will be on the operations of the CEGB.

The CEGB's general statutory duty was 'to develop and maintain an efficient, co-ordinated and economical system of supply of electricity'. Financially, the

Table 8.4   Public electricity generating capacity in the UK[a]

	CEGB	SSEB	NSHEB	NIE	Total	%
Hydro	2195	125	1764	–	4084	6.3
Nuclear	5069	1450	–	–	6519	10.1
Coal	30886	3888	–	240	35014	54.3
Oil	8417	–	1320[b]	1320	11057	17.1
Coal/oil	4504	–	–	–	4504	7.0
Coal/gas	366	–	–	–	366	0.6
Gas turbine	2517	55	35	240	2847	4.4
Other	–	–	146	–	146	0.2
Total	53954	5518	3265	1800	64537	100.0
Reserve plant	3141[c]	2028[d]	–	–	5169	8.0
Under construction	5101[e]	1400[e]	1[f]	360[g]	6862	10.6
Grand total	62196	8946	3266	2160	76568	118.6
Peak demand	46935	4125	1486	1261	53807	83.4

Notes: (a) at 31 March 1988; (b) being converted to natural gas firing; (c) oil/gas turbine plant; (d) oil-fired plant; (e) nuclear plant; (f) wind turbines; (g) coal-fired plant.

Sources: Handbook of Electricity Supply Statistics, Annual Reports of CEGB, SSEB, NSHEB, NIE.

ESI was obliged to remain financially in balance 'taking one year with another'.

However, in practice the powers of the CEGB have been severely circum-scribed by its relationship with government. The 1957 Electricity Act allowed the government:

> to give to the Electricity Council or to any of the Electricity Boards in England and Wales such directions of a general character as to the performance . . . of their functions as appear . . . to be requisite in the national interest.

Among the criticisms of the CEGB which subsequently emerged was that it was a conservative, engineering-led institution which has invested in unnecessarily large power plant. Since the early 1960s, the technological direction of the CEGB has remained fixed. Almost all of the fossil fuel-fired plant which has come on line has consisted of large units with a total site capacity of around 2000 MW. This policy gives the CEGB some of the largest plant in the Western world. Sixty per cent of its power output comes from only twelve large coal-fired power stations commissioned since 1966. This has had important implications for acid-rain control policy. Large power stations were chosen partly to minimize the difficulties of gaining planning consents for individual sites.

Coal for electricity supply

Ownership of the national grid and most of the country's power stations gives the CEGB, in principle, a great deal of flexibility in meeting electricity needs at the lowest possible cost. This flexibility was demonstrated during the 1984/85 miners' strike when system fuelling was changed round completely to minimize coal burn.

However, the CEGB has been bound in recent years by a 'Joint Understanding' with the British Coal Corporation (BCC) which has required it to burn some 75 m tonnes of UK coal a year. The Understanding, reached in 1979, initially offered the CEGB security of supply through a promise from the NCB to keep the price of coal constant in real terms.

Over time, increasing proportions of the coal supplied under the Joint Understanding have been bought at prices set according to the world market rather than BCC's own costs. In principle, the CEGB could have acquired imported coal at a far lower price for certain of its coastal power stations. The CEGB has argued that it could have displaced up to 30 m tonnes per year of British coal in this way.[5] However, this argument may be viewed as a negotiating stance, since purchases on such a large scale would inevitably have driven up the international price of coal.

The CEGB has imported small quantities of British coal in recent years (1 m tonnes in 1987/88) and also buys open-cast coal from UK producers other than BCC. An initial three-year agreement between the BCC and the private electricity generating companies will see the guaranteed volume of British coal sales fall to 65 m tonnes in 1992/93. Thereafter, coal imports could rise substantially.

Nuclear power in Britain

The experience of nuclear electricity generation has been, on the whole, an unhappy one. The CEGB was a reluctant convert to the cause of nuclear power in the 1950s, when it anticipated that costs would be higher than for coal-based electricity. The choice of reactor system has always been taken at the highest political levels. Until 1979, the belief that native nuclear expertise would allow Britain to develop its own designs of nuclear reactors with export potential meant that the world's leading reactor design developed in the US, the Pressurized Water Reactor (PWR), was shunned.

Britain's first nuclear design the the Magnox reactor. Nine of these, with a total output of 4000 MW, were commissioned between 1962 and 1971. Technically, their operation has been broadly satisfactory, although they have never generated power economically. The performance of Britain's second generation of nuclear plants, the AGRs, has been very disappointing. The first station, Dungeness B, was ordered in 1966. Only a fraction of its total planned capacity has been commissioned.

Four final AGRs have recently been commissioned. Three of these were ordered by the Labour Government in 1978, as part of a conscious decision to maintain an independent British nuclear design capability. There are indications that the performance of these stations might be more satisfactory.

At the same time as it ordered the three final AGRs, the Labour Government also decided to order one PWR in order to keep open the option of using standard world nuclear technology. One of the first acts of the incoming Conservative government in 1979 was to abandon the AGR option and announce plans for the construction of 15 PWRs by the end of the century. In the event, the planning process for the first of these, Sizewell B, took far longer than expected. After a lengthy Public Inquiry, planning permission was given only in March 1987. Already the estimated costs of Sizewell B have risen from £1.57bn to £1.79bn, breaching the initial contingency allowance. Plans for the construction of a further four PWRs by the end of the century were scrapped in November 1989 as electricity privatization exposed the true costs. A Public Inquiry into the first of these, Hinkley Point 'C', has lasted over a year.

Electricity supply and the environment

The nationalized electricity Boards had wide obligations, in statute and in practice, to promote the public and the national interest. When the 1957 Electricity Act was being drafted, environmental groups such as the Council for the Protection of Rural England (CPRE) lobbied strongly and successfully for the inclusion of specific environmental responsibilities to be laid on the Electricity Boards and the Government. The specific duty was:

In formulating or considering any proposals relating to the functions of the Generating Board or any of the Area Boards ... the Board in question, the Electricity Council and the Minister, having regard to the desirability of preserving natural beauty, of conserving flora, fauna and geological or physiographical features of special interest, and of protecting buildings and other objects of architectural or historic interest, shall each take into account any effect which the proposals could have on the natural beauty of the countryside or any such flora, fauna, features, buildings or objects.

The original objective in setting this requirement was to ensure the environmentally sympathetic construction of power stations and choice of sites. However, the duty has been interpreted, in an unintended manner, in the context of major environmental issues such as acid rain.

Environmental responsibility with the CEGB

At the time of its inception in 1957, the CEGB set up a Station Environment Group (SEG), reporting directly to the Board, within its Planning Department. The objective of the SEG was to help integrate environmental sensitivity into the planning of new power stations. The SEG was responsible for the much criticized 'tall stacks' policy practised in the 1950s and 1960s which improved air quality close to power stations but which helped to deposit emissions much further afield.

Increasingly, during the 1960s, the SEG began to take on a policy advice role as the importance of environmental issues from the point of view of public relations grew. In 1979, with a dearth of new power plant orders, the CEGB's Planning Department was wound up and the SEG transferred to the Technology Planning and Research Division (TPRD). In doing so, it lost its direct line of communication with the Board and became a small part of the CEGB's research activity.

Since the mid-1970s, when Scandinavian allegations about the consequences of the CEGB's sulphur dioxide emissions began to multiply, the TPRD research laboratories had been carrying out a significant programme of research into the long-range transport of air pollutants and their environmental impacts. Much high-quality scientific work was carried out. However, in the manner described in Chapter 3, the results tended to be deployed in a defensive way in order to highlight uncertainties and minimize the Board's apparent contribution to the acid-rain problem.

From 1979 onwards, the CEGB's posture on environmental issues in general, and acid rain in particular, was underpinned by a pure scientific perspective formulated in the most sceptical terms. Any counter-balance to the Board's scientific posture would have come from the old SEG which had been primarily concerned with power station design and pollution control.

By the early 1980s, the CEGB had become acutely aware that its environmental obligations under the Electricity Act were relevant to acid rain. This led to an exaggerated perception of the CEGB's role in formulating national environmental policies. The Chairman expressed the view in 1986 that 'the political challenge of dealing with acid rain or any other environmental problem caused by electricity production was met many years ago in the UK when Parliament wrote the relevant Acts which apply to the electricity industry', that the function of the pollution inspectorates was to act as a safety net 'if they consider that we are not doing our job properly' and that the government and Parliament are a further safety net in case both the CEGB and the inspectorates 'fail to win public confidence'.[6]

Thus, the CEGB's chairman asserted the virtues of self-regulation over the formal pollution control system, almost to the extent that any outside pressures to reduce emissions might be an interference in the CEGB's affairs.

The CEGB and flue gas desulphurization

By the early 1980s, a strong prejudice against the world's leading sulphur abatement technology, FGD, had emerged within the CEGB. Relatively little research and development work had been carried out and there was a belief that the effective clean-up of power station flue gases would have to wait until advanced coal combustion technologies were commercialized. To the extent that the CEGB was able to procrastinate about acid rain controls, this may have been a self-fulfilling prophecy.

The CEGB's predecessor companies were the first in the world to operate FGD. Units were fitted at three London power stations between 1930 and 1963. The original installations were at the request of the Commissioners of Works 'with a view to the protection of the royal palaces, parks and gardens, museums and other public buildings and their contents'. The FGD unit at Bankside, across the river from St Paul's Cathedral, operated with a lifetime SO_2 removal of 97 per cent until it was closed in 1979.

These FGD installations used sea water scrubbing, supplemented by additions of chalk, with effluent being discharged directly into the Thames. As the flue gases were not reheated after scrubbing, the chimney plumes were insufficiently buoyant and local ground level concentration of SO_2 remained high. The scrubber at Battersea in particular gave rise to more complaints from local authorities and the public than any other of the CEGB's stations.

Consequently, the CEGB's early experience with FGD was profoundly irritating. The Battersea scrubber, its digestor tanks lined with teak, was seen as little more than one step up from a laboratory prototype. The processes were troublesome and expensive to operate.

Thus, reports of the poor performance and availability of FGD installed in the US and Japan during the early 1970s were taken as confirmation that it was a technological dead-end. Even by the early 1980s, after satisfactory experience had been achieved abroad, this view still had currency within the CEGB. Greater faith was pinned in novel technologies such as fluidized bed combustion which, in the longer term, were expected to improve environmental performance and efficiency simultaneously.

The fact that this negative view went unchallenged may be attributed partly to the pattern of power plant ordering. It is far simpler and cheaper to fit FGD to a new power station than it is to retrofit an old one. As FGD technology began to mature in the late 1970s, no new fossil power stations were planned in the UK. The 2000 MW Drax extension, which began construction in 1978, had received approval from the government and the pollution inspectorate as early as 1965, long before modern FGD was commercial.

This early approval for Drax was, in turn, caused by the CEGB's policy of submitting planning applications well ahead of need in order to build up a stock of sites ready for development. The approval of 'anticipatory' planning applications allowed a lag to develop between the commercialization of new clean-air technology and its adoption in the UK. This lag covered the crucial period when FGD technology came to maturity and Britain failed to gain the technical expertise being accumulated in other countries.

Electricity supply in FR Germany

Structure

Given the complexity of electricity supply arrangements in the FR Germany, only the principal features of the system are described.

There are currently about 1000 public electricity companies, 600 of which are members of the Association of German Electricity Producers (Vereinigung Deutsche Electrizitätswerke–VDEW). VDEW members have a very wide range of characteristics – in 1987, 55 only generated electricity and 317 distributed power, while 310 performed both functions. In terms of numbers,

Table 8.5   Major electric utilities in FR Germany

Company	Home state/city	Electricity sales (TWh)	Generating capacity (GW)
DVG[1] Members			
RWE[2]	NRW[3]	119.1	27.4
Preussenelektra	Lower Saxony	51.8	12.8
Bayernwerk	Bavaria	32.3	7.0
VEW	NRW[3]	25.8	5.8
Badenwerk	Baden-Wurttemberg	16.9	5.0
EVS	Baden-Wurttemberg	16.8	4.5
HEW	Hamburg	11.8	4.1
BEWAG	Berlin	8.9	2.1
Non-DVG			
VEBA	NRW	15.8	4.3
STEAG	NRW	7.9	3.6
Others		47.8	10.1
Total Public Supply		354.9	86.7

Notes: (1) Deutsche Verbund Gesellschaft, the company owning the high-voltage transmission network; (2) also operates in Bavaria; (3) North-Rhine Westphalia.

the public electricity suppliers are dominated by municipally owned distribution companies which are organized into the separate Association of Municipal Undertakings (Verband Kommunaler Unternehmen–VKU).

Symbolizing the very distinct nature of the electricity supply industry and the special legal concessions allowed it, VDEW is not a member of the Association of German Industry (Bund der Deutschen Industrie–BDI). Given the diversity of its membership, VDEW cannot entirely resolve natural tensions between divergent interests. Although it can argue forcefully about issues such as the territorial concessions granted to the utilities, on topics such as environmental controls, which affect its members in different ways, it has been a less effective political force.

In considering the relationship between the electricity supply industry and government policy, the main interest lies in the eight companies which, through the German Grid Company (DVG), own the country's high-voltage transmission network. A profile of the eight members, plus two other major utilities, is given in Table 8.5.

Table 8.6 shows that there are wide variations between the DVG members in terms of the types of generating capacity employed. In particular, Germany's leading utility, Rheinisch-Westfälisches Elektrizitätswerk (RWE), relies on brown coal for a large proportion of its fuel supplies. RWE also generates more nuclear power than any other utility, giving it the advantage of the two cheapest forms of electrical generation available in FR Germany. Hard coal is the dominant fuel throughout the utility system. However, Bayernwerk, the Bavarian utility, breaks the pattern, being reliant primarily on hydro-electricity, nuclear and oil.

Table 8.6 Public electricity generating capacity in FR Germany

	Total	%	RWE	– of which Preussenelektra	Bayernwerk
Hydro	6204	7.2	2753	894	1810
Nuclear	19778	22.8	5768	1970	1952
Brown Coal	12775	14.7	11179	1208	–
Hard Coal[1]	26560	30.6	966	2991	800
Gas	12097	14.0	600	1270	–
Oil[2]	8694	10.0	1428	–	1825
Other[3]	565	0.7	4698	4517	597
Total	86673	100.0	27392	12850	6984

Notes: (1) includes coal/oil or coal/gas firing; (2) includes oil/gas firing; (3) for individual utilities, includes small plant, supplies from subsidiary companies and contracted external power supplies.
Source: Derived from Statistik für das Jahr 1987 (VDEW).

Between 15 and 25 per cent of the share capital in the largest utilities is in public ownership, principally through the Länder and municipalities. The most interesting example of public participation is RWE. While the majority of shares are in the private sector, the municipalities enjoy greater influence because their shares account for the majority of votes.

The utilities have become an important source of revenue for many Länder and municipalities, which tempers the desire to take measures which might adversely affect profits. For this reason, the Länder and the Bundesrat (see Chapter 6) have a particularly strong interest in utility policy.

The German utility sector is characterized by complex patterns of cross-ownership linking utilities with each other as well as with sources of fuel supply. It is possible to give only a few of the examples which abound. There are also many shared directorships between different companies.

RWE owns almost all of the shares in Rheinische Braunkohlenwerke (RBK), which supplies it with all its brown coal, and substantial shares in the nuclear fuel cycle. RWE acquired its ownership of brown coal deposits when Jewish business interests were appropriated during the 1930s, a fact which partially explains Green antipathy to the company. RWE also has an indirect stake in the second largest German utility, VEW. Recently, RWE acquired construction and waste management interests in order to gain some control over the markets for gypsum produced in its FGD units.

The major energy conglomerate VEBA, the origins of which are the combined coal and electricity interests of the pre-War Prussian State, has a major stake in Ruhrkohle and also owns the three utilities Preussenelektra, Nordwestdeutsche Kraftwerke (NWK) and VEBA Kraftwerke Ruhr. These electricity companies form a major market for Ruhrkohle's output. In turn, Ruhrkohle has a majority stake in Steinkohlen Elektrizitäts AG (STEAG), an electricity generating company founded by the mining companies in 1937.

History of electricity supply

The reasons for the pluralistic and convoluted structure of the ESI lie in German history and the complex set of relationship between industry and the state.

The great utilities of today developed by challenging the power of the municipalities and the mine owners who were the first electricity generators. A separate development was the exploitation of hydro-electric power in Southern Germany. RWE's first power station used steam supplied by an enterprising mine-owner circumventing the production restrictions of the Ruhr Coal Syndicate. Gradually, companies such as RWE began to generate electricity at the mine-mouth and supply it to the municipalities. In doing so, suspicion about the potential powers of a privately owned monopoly was invoked.

The municipalities had the counter-balancing power of being able to grant wayleaves for transmission lines. A compromise was established whereby the utilities' territories were marked out, while the municipalities took major share-holdings. Deals were also struck with the mine-owners, who agreed to forego the right to build transmission lines connecting different mines in exchange for electricity supply contracts.

The current pattern was largely fixed by the 1935 Energy Industry Act (EnWiG). This ended a period of difficulty and internal conflict for the electricity industry. There had, before then, been a debate about state ownership versus private ownership under state supervision. The EnWiG represented a compromise by which the industry avoided nationalization while remaining subject to official guidance in order to satisfy the demands of the state for cheap, secure and abundant energy. Economic regulation of the industry takes place at the Land rather than the Federal level.

The question of national ownership also arose during the post-war Allied occupation. However, the utilities resisted and the Allies were reluctant to see the creation of such a substantial concentration of economic power. The provisions of the EnWiG were thus carried over intact when the Federal Republic was created in 1949.

The energy industry act and its effects

The EnWiG provides territorial concessions to all public electricity companies and, importantly, exempts them from anti-trust legislation. Concessionary contracts with municipalities give utilities the exclusive right to supply power in that area. At the same time, demarcation contracts between utilities ensure that one utility does not encroach another's territory. In exchange for these rights, the utilities are required to provide cheap and reliable power and must have their tariffs approved by the Federal and Land governments. Electricity prices are set on the 'cost-plus' principle, allowing the utilities a reasonable return on capital employed.

In addition, companies must make their transmission lines available to other public electicity suppliers on reasonable terms. However, these rights are not extended to third parties, so that large industrial conglomerates may experience difficulties in trying to transfer power from one site to another over the transmission network.

The EnWiG creates a framework within which the public electricity suppliers co-operate and complete within defined limits. While co-operation takes place over territorial demarcations and over transfers of power to ensure security of supply, there is no country-wide system of optimal power station dispatch on an hour-to-hour basis such as that practised in the UK.

On the other hand, bulk transfers of power, notably from North to South, do take place, as does third party use of transmission networks within the public electricity sector. Competition for licences to supply smaller distribution companies is beginning to emerge. Equally, although utilities cannot compete within the same territory, there is competition to attract major industrial consumers to a particular geographical location by offering attractive power prices. Thus, a utility can expand its market share without encroaching another's territory.

Those provisions of the EnWiG which are anti-competitive in nature are attracting considerable criticism from large industrial consumers, municipalities and the European Commission, particularly in the context of the 1992 objectives. The EnWiG is also a useful stick with which the green movement can beat the utilities, given that it tends to discourage small-scale dispersed electricity generation.

In the early 1980s, both the Federal and Länder governments were beginning to seek more power over the electricity sector. For example, all the concessionary contracts, which provide the utilities with legally protected supply areas, will have to be renegotiated by 1994. In addition, price increases requested by the utilities have not been as readily approved in recent years as in the past. This symbolizes an official objective to introduce price competition between oligopolistic companies and between the different sources of fuel supply for electricity generation.

Environmental regulations proved helpful to the Federal government as it sought influence in the 1970s and early 1980s. Investment was squeezed out of the utilities, large-scale electricity generation was favoured, the effective costs of both brown and hard coal were raised, and government gained access to more financial information about the utilities' activities.

Coal supply for electricity generation

RWE is responsible for most of the consumption and production of brown coal through its wholly owned subsidiary Rheinische Braunkohlenwerke. Until the acid rain controls which were put in place during the 1980s took effect, electricity generated from brown coal was the cheapest in FR Germany. According to the industry, this position is now occupied by nuclear power.

Hard coal supplies, being tied closely to the fortunes of the German mining industry, are politically problematical. As in Britain, the use of domestically produced coal for electricity generation is higher than it would be if costs alone were allowed to dictate the outcome. Given the complex utility structure in

Germany, and the legally defined powers of the Länder, the economic instruments required are more formal than the CEGB/BCC 'Joint Understanding'.

The first of two devices used is the Jahrhundertvertrag ('century contract') established between Ruhrkohle and the VDEW acting on behalf of the major utilities. It was first signed in 1977 at the behest of the Federal government as part of its response to the oil crisis. The utilities accepted the Jahrhundertvertrag only under duress – legislation was threatened if a voluntary agreement with the coal producers was not reached.

The initial contract lasted between 1978 and 1987 and required an average offtake of 33 m tonnes per year. In 1981, the agreement was extended to 1995. The annual offtakes were set to rise to 47.5 m tonnes in 1995. This latter target was reduced to 41 m tonnes/year in 1989. The Jahrhundertvertrag, coupled with rising nuclear output and a slowdown in the rate of increase of electricity demand are, as Figure 8.1 shows, beginning to squeeze the use of inexpensive brown coal.

VDEW won a concession that the additional costs of burning domestic coal should be transparent to customers and should not be 'lost' in their tariffs. In order to achieve this objective, the Kohlepfennig ('coal penny') was devised. The system works as follows. The BMWi sets a price at which coal is traded between the coal companies and the utilities. Those utilities burning hard coal are compensated for the amount by which the price of hard coal exceeds the price of imported coal or the price of oil. About a third of hard coal use is subsidized at the domestic/imported coal price differential and the remainder at the coal/oil price differential. Prior to 1986, when oil prices were high, the second part of the subsidy was not operative.

The Kohlepfennig fund is maintained by a levy on all electricity sold in the FR Germany and appears as an explicit item on all electricity bills. There are regional variations in the amount charged, but the average is currently 8.5 per cent. Since the Kohlepfennig is a transfer mechanism internal to the electricity supply industry, the Federal government now argues that it is not a state aid within the terms of the Treaty of Rome.[7]

The Kohlepfennig is resented in Southern Germany where there is a low reliance on coal for electricity generation and the Bavarian-based CSU objected to its introduction in the 1970s. In 1984, a challenge to the constitutionality of the Kohlepfennig failed in the Federal Administrative Court. Southern resentment redoubled in 1986 as the degree of subsidy soared when the world price of oil fell well below that of German coal. Ironically, this resulted in benefits for utilities and consumers in NRW where unprofitable coal is mined and used, but also in price increases in Bavaria thanks to the increased level of subsidy.

Regional tensions, particularly North–South ones, over the Kohlepfennig also contributed to the Waldsterben debate in 1982–3. As will be described in Chapter 10, tighter environmental controls which would increase generation costs in the North were seen, by CSU politicians, as *juste retour* for forest damage in Bavaria and the Kohlepfennig subsidy, as well as an opportunity to enhance the prospects of an increase in the use of nuclear power.

Nuclear power in Germany

The origins of nuclear power in Germany lay in the desire, back in the 1950s, to build a high-technology industry which could be used as an engine for economic recovery and export opportunities. In this respect, there are echoes of the later concern with environmental protection technology. In 1955, the Federal Ministry for Atomic Affairs was established, headed by Franz-Josef Strauss, later the strongly pro-nuclear Prime Minister of Bavaria at the time when Waldsterben was to become a major policy issue. Since the mid-1960s more than 20 per cent of the Federal science budget has been allocated to nuclear R&D, but this has now declined.

Two German companies originally held licences for the manufacture of US reactor designs, AEG for boiling water reactors (BWRs) and Siemens for PWRs. As in the UK, the utilities were initially sceptical about the economic benefits of nuclear power.

In 1969, Siemens and AEG merged their nuclear production facilities in Kraftwerk Union (KWU) which has built all but one of the subsequent German reactors. A total of 23 nuclear plants with a total capacity of almost 20 000 MW now operate in the country. While this is considerably less than the 45–50 000 MW planned for 1985 in the immediate aftermath of the 1973 oil crisis, it is still a remarkable technical achievement.

The comparative success of the nuclear programme is due primarily to the efforts of RWE supported by the Federal Ministry for Research and Technology (Bundesministerium für Forschung und Technologie, BMFT). RWE has consolidated its dominance of the German utility industry through its technical expertise as an informed buyer of reactor technology. RWE's leadership stems from the fact that it is the only utility operating a supply system large enough to absorb a single large nuclear reactor. Other German utilities operating smaller systems have had to pool their resources and share the output of individual reactors. RWE pioneered the convoy concept whereby a group of technically identical reactors are pushed into the licensing process simultaneously, allowing returns to scale in the manufacture of components. Construction of these began in 1982.

As described in Chapter 5, the nuclear industry has been subject to concerted political opposition. The Bürgerinitiativen have repeatedly challenged nuclear power plant construction in the courts. While utilities have often been able to continue the construction of plants while lengthy court procedures were followed, they have faced the risk of carrying out expensive retrofits before being allowed to begin producing power. By the end of the 1970s, lengthy licensing procedures had effectively halted the construction of new nuclear power plants.

The other potential obstacle to nucler power is the question of fuel reprocessing and disposal. An amendment to the Atom Gesetz (Nuclear Act) of 1976 makes utilities responsible for spent fuel once it has been removed from the nuclear reactor. The ideas of storing, handling, reprocessing and disposing of spent fuel are encapsulated in the single German word *Entsorgung*. The only part of the Entsorgung process excluded from utility responsibility and allocated to the Federal government is the ultimate storage of radioactive waste.

Utilities must demonstrate that they can execute the Entsorgung process before the courts may grant them licences to build a nuclear plant. Safety and security, unqualified by economic considerations, are the only criteria for determining Entsorgung.

A depository for the final storage of waste is still under evaluation at Gorleben in Lower Saxony. Construction of the reprocessing plant at Wackersdorf in Bavaria, which had already begun, was abandoned because of political resistance and the high costs faced by the utilities early in 1989. The utilities will now have their fuel reprocessed in the France and the UK.

The fact that the SPD has decided not to allow the construction of any more nuclear plants should it be elected back into office, coupled with the even greater public mistrust of nuclear power which has emerged following the Chernobyl accident, suggests that the prospects for future nuclear construction are bleak for a long time to come.

Conclusions

While there are certain strong similarities between the energy sectors in the UK and in the FR Germany, there are also major differences, particularly in terms of the ownership of major companies, industrial structure and regulation. Both countries are highly dependent on domestically produced coal as a power station fuel and have faced major economic and social problems associated with high cost production. While both have attempted to promote nuclear power, the FR Germany has been notably more successful in the attempt in spite of strong public opposition.

From the point of view of acid-rain controls, arrangements for electricity supply are the most important factor distinguishing the two countries. In Britain, the nationalized ESI has been very tightly bound to central government and many strategic decisions have been taken at a political level. Conversely, the CEGB has been able to play a major role in determining policy on acid emissions. Electricity privatization will completely·change the nature of these relationships.

On the other hand, the more numerous German utilities have a more complex relationship with government at the Federal, Land and municipal levels, through both regulation, and ownership. Perceived regional injustices in the system of financial aid used to subsidize the German hard coal industry exchanges (the Kohlepfennig) were a factor encouraging environmental controls which were seen to weaken coal's market position. In Britain, environmental arguments against coal utilization began to be advanced only later in the 1980s in the context of the global climate change debate.

Notes

1.   This chapter has been compiled from the authors' knowledge and from a variety of sources, including:

J.H. Chesshire and J.F. Skea (1989), *The UK Energy Market*, James Capel & Co, London.

T. Daintith and L. Hancher (eds) (1986), *Energy Strategy in Europe: The Legal Framework*, de Gruyter, Berlin.

J. Franke and D. Viefhus (1983), *Das Ende des billigen Atomstroms*, Verlag Kölner Volksblat.

T. Hofer-Busse (1985), *RWE-Ein Riese mit Austrahlung*, Verlag Kölner Volksblatt.

F.J. Holker and F. Raudszus (1985), *Die Konzentration der Energiewirtschaft- Kritik der Ordnungspolitik im Energiesektor der BRD*, Campus, Frankfurt am Main.

S. Köhler (April 1984), *Geschichte der deutschen Energiewirtschaft und ihre Auswirkung auf die kommunale und regionale Energieversorgung*, Öko-Insitut Werkstattreihe.

N. Lucas (1985), *Western European Energy Policies*, Clarendon, Oxford.

L. Mez (1984), *Neue Wege in der Luftreinhaltepolitik: Eine Fallstudie zum informalen Verwaltungshandeln in der Umweltpolitik am Beispiel RWE*, IIUG Report 84–3, Wissenschaftszentrum, Berlin.

J. Radkau (1980), *Aufstieg und Krise der Deutschen Atomwirtschaft*, Rowohlt.

D. Rehfeld (1986), *Bestimmungsfaktoren der Energiepolitik in der Bundesrepublik Deutschland*, Lang, Frankfurt am Main.

H-W. Schiffer (1985), *Structur und Wandel der Energiewirtschaft in der BRD*, Verlag TUV Rheinland.

J. Strasser and K. Traube (1981), *Die Zukunft des Fortschritts*, Neue Gesellschaft, Bonn.

S.D. Thomas (1988), *The Realities of Nuclear Power: International Economic and Regulatory Experience*, Cambridge University Press, Cambridge.

2. Nuclear power, dependent on imported uranium, is not viewed as an indigenous source of energy by the environmental movement. However, energy suppliers focus more on the relatively low contribution of uranium supplies to total nuclear costs and to the lack of any short-term vulnerability to supply interruptions.

3. International Energy Agency (1989), *Energy Policies and Programmes in IEA Countries 1988*, OECD, Paris.

4. Coal Industry Tripartite Group (1977), *Progress with the Plan for Coal*, National Coal Board, London.

5. House of Commons Energy Committee (1986), *The Coal Industry: Memoranda Laid before the Energy Committee*, HC 196–I, Session 1985–86, HMSO, London.

6. Lord Marshall of Goring (1986), An industry viewpoint, Proceedings of the NSCA/IEEP conference, *Acid Rain: The Political Challenge*, 19 September 1986, London.

7. The European Commission disagrees and has demanded that Germany present a plan for restructuring the coal industry and reducing coal subsidies by 1993. Germany, in turn, has presented a case repudiating the Commission's demand to the European Court of Justice. The Commission has threatened not to approve Kohlepfennig payments beyond 1990. See *Power in Europe*, No 70, 29 March 1990, FTBI, London.

9 The control of air pollution

Introduction

The purpose of this chapter is to explain how, and to what extent, the laws and the instutitions relating to air-pollution control affected the wider political process of policy formation in Britain and Germany. The formal framework for the regulation of atmospheric emissions during the acid-rain debate in each country is described in some detail and its role, if any, in determining policy is analysed.

The chapter divides into three sections: the historical development of air-pollution control regimes in each country; the legislative frameworks, including a discussion of underlying control principles and concepts; and implementation regimes. The space devoted to each of these differs for each country, reflecting fundamental differences in the style of regulation.

It will become clear that policy choices are severely constrained by these legal and institutional factors. The institutions responsible for air pollution control were central to the determination of national positions only in FR Germany.

In Germany, environmental legislation is both recent and conceptually ambitious, with the courts playing a greater role in defining and clarifying the basic objectives and regulatory instruments. Formal legislative provisions in Germany receive more attention in this chapter than do those in Britain, where the regulatory authorities have enjoyed greater discretion under legislation which has accumulated over a century and which is drafted in terms of general requirements. Consequently, for Britain, more attention is focused on implementation procedures.

The chapter explains those aspects of the pollution control regimes which affected the acid-rain debate during the early to mid-1980s. More recent developments are reviewed in Chapter 13. For Germany, the focus is on the legal position up to 1983 when the Large Combustion Plant Regulation (Grossfeurungsanlagenverordnung-GFAVo) came into force. However, somewhat later decisions on vehicle emissions and smaller pollution sources are mentioned briefly. In Britain, where changes have been stimulated not by domestic pressures but by European obligations, 1986–7 is taken as the break point.

The development of air pollution control

In both Britain and Germany, formal provisions for air pollution control go back some considerable time. The earliest records in Britain date back to the thirteenth century, when Edward I prohibited the burning of sea-coal in London. The problem of polluted air in the English capital was to remain an issue until well into the twentieth century. The first instances of air pollution controls in Germany date from the early eighteenth century when Saxon princes issued regulations restricting emissions from metal smelting industries.

In both countries the foundations of the present air-pollution control regime were laid in the middle of the nineteenth century and were stimulated by the air-pollution impacts of industrialization. In Britain, there has been a remarkable degree of continuity between the nineteenth century and the present. In Germany, the legislative structure is, on the whole, comparatively recent. Developments in each of the countries are examined in turn.

Air pollution control in Britain[1]

In 1862, *The Times* reported that 'whole tracts of country . . . have been swept by deadly blights till they are as barren as the shores of the Dead Sea'. This environmental damage was attributed to the uncontrolled emissions of hydrochloric acid from the growing chemicals industry.

With the support of vociferous agricultural interests, Parliament passed the 1863 Alkali Act which established an 'Alkali Inspectorate' and created powers to impose substantial fines on factory owners who failed to reduce their emissions by 95 per cent. The Alkali Inspectorate was intended to be independent of local authorities in which factory owners often occupied influential positions.

A distinctive feature of the Alkali Inspectorate's early methods was the emphasis on the co-operative, as opposed to an adversarial, approach to dealings with registered works. Factory owners had it within their power to make the Inspectors' lives highly uncomfortable, not to mention positively dangerous. This created an early motivation for accommodation and negotiation with manufacturers, which was later to become an article of faith for the Inspectorate. Robert Angus Smith (the first Alkali Inspector and the coiner, in 1872, of the term 'acid rain') made himself popular with works owners by demonstrating how condensed hydrochloric acid could be profitably recycled and the Inspectors soon began to adopt the role of free technical consultants to the chemical industry.

Another factor which pushed the Inspectors closer to manufacturers was the lack of appreciation of their work from the government departments to which they reported. Sensitive to continuing pressure from agricultural interests, the Local Government Board at one point linked salary increases for the Alkali Inspectors to an evident increase in air quality and the rate of prosecution of factory owners.

The provisions of the first Alkali Act were extended indefinitely in 1868 and statutory emission limits (the last until 1990) were added in 1874. Finally, the

1906 Alkali &c Works Regulation Act sealed the pattern of regulation of industrial air pollution from major sources for most of the twentieth century. This included a schedule of 'noxious and offensive gases' to be regulated, a similar list of scheduled industrial process and it confirmed that the principle of 'best practicable means' (BPM) should be used to regulate emissions where no statutory limits existed.

THE SMOKE PROBLEM AND THE CLEAN AIR ACT

There had been attempts since the early nineteenth century to deal with the problem of smoke pollution in Britain at a national level. These attempts failed largely because of the reluctance of manufacturing industry to bear the costs of introducing new combustion equipment, a national affection for the open fire and a culturally based resentment at the prospect of the government's having jurisdiction over what people did in their own homes.[2]

In 1952, public attitudes to smoke control changed completely. Until then, the smoke problem had been perceived almost entirely in terms of 'nuisances'. However, the 4000 excess deaths from respiratory and heart disease which were attributed to London's great fog in December 1952 brought a sense of national urgency to the smoke control problem.

In spite of the degree of concern, government ministers adopted the same rhetoric as did their successors with respect to the acid rain issue and declined at first to take any action, citing, among other factors, the economic implications. However, the Beaver Committee which was appointed to assess the problem recommended significant reductions in emissions of smoke, particulate matter (grit and dust) and sulphur dioxide (SO_2). It proposed the passing of a new Clean Air Act[3] requiring: (1) major restrictions on emissions of smoke, grit and dust from industrial furnaces; (2) that industrial processes presenting special technical difficulties, including electricity generation, should be entrusted to the Alkali Inspectorate; (3) that local authorities should be empowered to set up 'smokeless zones; and (4) that local and central government should contribute to the cost of converting household appliances to burn smokeless fuels.

In addition, it was suggested that flue gas desulphurization (FGD) should be fitted at all new power stations in or near populated areas. This suggestion was not taken up by government given the cost and the unsatisfactory nature of the FGD processes (described in Chapter 8) already operating in the UK.

The government decided 'in principle to adopt the policy recommended by the Beaver Committee' and was forced to put this policy into practice quickly when it was embarrassed by a back-bench Member of Parliament who introduced a private Clean Air (Anti-Smog) Bill only days after the publication of the Beaver Report. The 1956 Clean Air Act, while falling somewhat short of the Beaver recommendations, represented a major step forward in the control of air pollution in the UK.

In the years that followed, significant improvements took place in British air quality, particularly in urban areas. In reality, much of this can be attributed to factors other than the passage of the Clean Air Act, including the replacement of coal burning in open fires by gas and electric central heating and the flight from inner city areas towards the suburbs.[4] Nevertheless, the Act has had an enormous symbolic significance and has been the focus of much admiration, including in the FR Germany. However, the apparent success of the Act contributed to the development of an unwarranted complacency which hindered effective action when new environmental problems arose.

Major processes and the Clean Air Act

Using powers granted under the 1956 Clean Air Act, several new industrial processes were removed from the jurisdiction of local authorities and transferred to the Alkali Inspectorate under the 1958 Alkali &c Works Order. This followed a nine-day statutory public inquiry which was characterized by a scramble by various industries for the scheduling of their processes. Many industries felt unable to comply with the basic requirements of the Clean Air Act and believed that, if they succeeded in making the case that emissions from their plants were technically difficult to control, the Alkali Inspectors would take a sympathetic view of the problems which they faced. Industry's expectation that regulation at the national level would protect it from what were thought to be excessive local demands would be echoed later in FR Germany with respect to power plant emissions.

Notably, while the schedule of noxious substances attached to the 1956 Clean Air Act included SO_2, it specifically excluded SO_2 deriving exclusively from the combustion of coal, as had been the case since the 1906 Alkali Act. On a formal basis, this kept power stations SO_2 emissions out of the jurisdiction of the Alkali Inspectorate. However, the Central Electricity Generating Board (CEGB) and the Inspectorate agreed informally to proceed as though power station SO_2 was a scheduled substance.

AIR POLLUTION AND THE WORKPLACE

From 1975 to 1987 the Alkali Inspectorate – by then known as the Alkali and Clean Air Inspectorate (ACAI) and renamed the Industrial Air Pollution Inspectorate (IAPI) in 1982 – formed part of the Health and Safety Executive (HSE) which has primary responsibility for regulating conditions within the place of work. This arrangement had major repercussions for the status and morale of the Inspectorate during the late 1970s and 1980s when acid rain became a political issue.

The Inspectorate was incorporated into the HSE under the 1974 Health and Safety at Work (HSW) Act. The intention was to rationalize industrial inspection and to remove confusion about the respective roles of the Alkali Inspectorate and the Factory Inspectorate when dealing with serious pollution incidents by merging the two bodies.

The Inspectorate was demoralized by this apparent end to its independent existence. However, no unification with the Factory Inspectorate took place. Just before the 1974 HSW Act received the Royal assent, the government invited the Royal Commission on Environmental Pollution (RCEP) to undertake a review of 'the efficacy of the methods of control of air pollution from domestic and industrial sources, to consider the relationship between the relevant authorities and to make recommendations'.[5] In order not to prejudge the findings of the RCEP inquiry, no irreversible steps, such as merging the pollution and factory inspection functions, were taken.

The RCEP's 5th Report in 1976 criticized the ACAI's move to the HSE in unequivocal terms. It recommended that a unified pollution inspectorate operating within the Department of the Environment (DoE) be set up, using a new principle (best practicable environmental option – BPEO) to balance the impacts of emissions to land, air and water.

Their objections to the ACAI within HSE were that (1) environmental pollution relates to amenity and property damage as well as health and safety; (2) the criteria for acceptable exposure to pollution are different in and outside the place of work; (3) the Health and Safety Commission (HSC) was too dominated by industrial interests and was too narrowly based to reflect the broader public interest; and (4) environmental concerns would be lost in an organization such as HSE. This last point is borne out by a reported remark made by the Director-General of HSE to IAPI inspectors in 1984, describing them as 'a pimple on the face of HSE'.[6]

In its belated 1982 reply, the government rejected the recommendations in the RCEP's 5th Report, expressing scepticism that the proposals made would improve control arrangements.[7] However, a major review of air pollution was promised. This was concluded only in 1986, ten years after the publication of the RCEP's original report.

In 1982, a memorandum of understanding was drawn up in order to clarify the lines of responsibility for air pollution control.[8] The HSE were to carry out their responsibilities 'in the light of general policies of pollution control established by the Environment Ministers' and it committed itself to maintaining a separate group of industrial air pollution inspectors.

The ultimate consequences of the government's re-assessment of air-pollution control, including the establishment of a unified Inspectorate of Pollution and the effective adoption of BPEO, in the guise of integrated pollution control (IPC), are described in Chapter 13.

Thus, in spite of the HSW Act, the existence of the Inspectorate was guaranteed and pollution control arrangements could continue much as before, although staff remained deeply unhappy about the institutional framework within which they operated. Institutional wrangling, political neglect and isolation from the policy process has characterized the Inspectorate during the 1980s. It is difficult to conceive of pro-active measures to restrict acid emissions originating from the Inspectorate against this background.

Air pollution control in West Germany[9]

Air pollution control became a component of the German legal system when, under French influence, the Prussian state issued General Trade Regulations (Gewerbeordnung-GeWO) in 1845. In 1869, this law was adopted by the first German Reich. It empowered the bureaucracy and police, through numerous trade offices (Gewerbeaufsichtsämter), to restrict private property rights in the public interest.[10]

Laws were therefore implemented, as they are today, as part of the general plant authorisation procedures by local or regional officials who are guided by administrative directives (Verwaltungsvorschriften) and associated technical guidelines (Technische Anleitungen – TA) or are instructed by binding regulations. Many requirements other than environmental controls are laid down during these highly formalized authorization procedures.

Directives aimed at improving health and safety standards flourished briefly prior to the First World War. However, subsequent political and economic crises led to neglect and poor implementation. While the preoccupation with 'Die Natur' during the Third Reich might have been expected to enhance the priority attached to what is now called environmental regulation, very little was achieved except in the realm of nature conservation.

POLLUTION CONTROL DURING RECONSTRUCTION

Germany emerged from the Second World War with little in the way of environmental controls apart from the nineteenth-century GeWO provisions which gave local authorities limited powers. Under the GeWO, emissions from a plant had had to be tolerated as long as these were 'customary in the location' (ortsüblich). Abatement measures proposed also had to be economically defensible. In practice, this meant that the emissions from a particular plant (as opposed to the general level of air quality) had to be demonstrated to be causing ill health among the public.

The salience of environmental issues during reconstruction was understandably low, except in the Land of North Rhine-Westphalia (NRW) where industry and people were crowded together in Germany's traditional industrial heartland. The prevailing emphasis on free-market ideology was not conducive to the growth of state regulation, especially at the Federal level.

In NRW the first battles over air pollution took place during the 1950s, with the British Clean Air Act as a model. Proposals to improve air quality initially came from municipalities which wanted to set up a regulatory system similar to that used for water pollution control, based largely on citizen organizations rather than regional or local government. The Federation of German Industry (BDI) strongly opposed these proposals and, in the end, agreed only to a few measures which slightly increased the powers of the trade offices, extended the list of plants requiring permits and updated the technical guidelines associated with the GeWO.

Because of its early interest in air pollution control, and because of the heavy concentration of power stations and other polluting industries within its boundaries, NRW became the region with the greatest experience of air-pollution control.

In 1952, an Interparliamentary Working Group for Nature-Appropriate Economic Activities (Interparliamentarische Arbeitsgemeinschaft für naturgemmässe Wirtschaftsweise – IPA), an all-party committee of MPs, was set up. It became increasingly concerned about poor air quality and subscribed to principles which now sound astonishingly modern: concern about physical limits to economic growth, man's moral responsibility for the Earth, the significance of non-economic values and the need for ecological understanding.

It became the major challenger to the BDI on environmental issues during the 1960s. Its influence remained confined, however, to the Federal administration.

THE 1959 FEDERAL AIR PURITY ACT

The pressure for national legislation, following the British and NRW examples, was strong enough to create, in 1959, a Federal Air Purity Act (Luftreinhalte-gesetz – LRG) which extended GeWO provisions. This first legal change since 1899 contained little more than that which the BDI had already conceded in NRW. The primary instruments to be used were ambient air-quality standards.

The LRG weakened somewhat the requirement to demonstrate public health impacts in order to justify abatement, extended the list of plants requiring permits, modernized the technical guidelines associated with the granting of authorizations and, importantly, broadened the application of the principle of 'economic defensibility' to apply to sectors of industry rather than to individual firms. Domestic and mobile sources were not covered by the LRG, largely because of the vociferous opposition of the oil and gas industries. Not surprisingly, oil and gas became the first targets of Federal environmental regulation in the 1970s.

Licensing authorities were also empowered, for the first time, to require the new emitters be equipped with state-of-the-art (Stand der Technik) emission control technology. What this meant in practical terms was left to the judgement of the Länder authorities and the courts.

Soon, environmental groups were arguing that the prevailing control regime based on air-quality criteria was not working. This motivated politicians to take up the issue and in 1961 Willy Brandt made 'blue skies over the Ruhr' one of the SPD's slogans for the Federal elections. In 1961, the setting up of the Federal Health Ministry (Bundesministerium für Gesundheitswesen – BMG), then with responsibility for air-pollution policy, marked the first milestone on the road towards Federal control.

It took until 1964 to implement the LRG because of wrangling over the associated technical guidelines. The BDI succeeded in having the 'Stand der Technik' criterion qualified by the traditional concept of 'economic defensibility'.

Environmental policy in West Germany during the mid-1960s was therefore as 'reactive' as Britain's a decade or so later. However, the position changed in 1969 when an SPD/FDP coalition government was formed. Germany was swept by a 'reform euphoria' based to some extent on views and ideals copied from the North American environmental movement.[11] As noted in Chapter 4, environmental issues struck a deeper chord in German society than in Britain. It was at this time that the Federal Interior Ministry (Bundesministerium des Innern-BMI) acquired the primary responsibility for air pollution from the BMG. After 1969, controlling air pollution in order to protect first human health and later the natural environment became a core concern first of the Federal bureaucracy and then the politicians.

THE 1974 FEDERAL AIR QUALITY ACT

Following the 1972 Stockholm Conference on the Human Environment, the new concept of environmental protection – Umweltschutz – was included in a new Article to the Constitution. This enabled the Federal government to legislate in new areas and effectively removed much environmental policy-making from the Länder to Bonn. In particular, concurrent legislative competence for air, water, waste and noise were added to the existing Federal responsibilities for nuclear safety and radioactive waste management. The Vorsorge (precautionary) principle was specifically included as a cornerstone of environmental policy. The regulation of emissions from fossil fuels became the responsibility of one Ministry, the BMI.

This change in the Constitution enabled the Federal government to begin translating a conceptually very progressive environmental programme into legislation. This was encouraged at the very top by a small number of senior politicans including the Chancellor, Willy Brandt, and the Interior Minister, the FDP's Hans-Dietrich Genscher. In 1974, the GeWO was replaced by the Federal Air Quality Protection Act (Bundes-immissionsschutzgesetz – BImSchG), which potentially covers all sources of air pollution.

The BImSchG brought air-pollution control firmly into the Federal domain and covered the protection of fauna, flora, water and soil as well as human health. The BImSchG was based on the 1955 Atom Law, with one important difference. While nuclear safety was to be achieved by technological and scientific means without regard to cost, economic defensibility remained a component of the BImSchG, but only for retrospective requirements applied to existing plants.

There had been considerable initial opposition to the new act, especially from the Southern Länder, which had not been active in furthering air pollution abatement and from NRW, all of which were at the time ruled by the CDU/CSU. These Länder saw no need for the BImSchG, felt unable to implement it and were fundamentally opposed to national uniformity.

The BImSchG enabled the Federal Government, through the BMI, to pass directly binding regulations (Verordnungen) as well as administrative directives, a task which it carried out with enthusiasm, tactical skill and considerable philosophical thoroughness. Its Secretary of State responsible for environmental protection until 1982, Dr Gunther Hartkopf, later described these early years as consisting primarily of battles against industry and its supporters.[12] The BMI was aided by expert advice from the newly created Federal Environment Office which acts as a source of broadly based professional advice to the BMI, industry and the public.

UBA represents an important repository of expertise and environmental enthusiasm which has helped to provide policy continuity even when environ-mental issues had a low political priority. It co-ordinates environmental policy implementation between the Länder and fulfils an important role in educating and informing the public about environmental issues. Its staff, now over 500, includes scientists, engineers, economists, other social scientists and lawyers. This disciplinary mix ensures that government can take a broad view of the impacts of new environmental measures.

The BImSchG was only one result of the very ambitions environment programme adopted by the SPD/FDP government in the early 1970s. However, the inability of the SPD/FDP coalition to implement fully this programme contributed to the disenchantment of voters and gave rise to the perception of an implementation gap (Vollzugsdefizit), much studied by German social scientists, including those reporting to the government.[13] As described in Chapter 5, the Vollzugsdefizit in turn provided ammunition for the growing green movement.

The legislative framework

Legislation in Britain

The 1990 Environmental Protection Act has significantly changed the legislative framework for air pollution control in the UK. This is briefly described in Chapter 13. However, the acid-rain issue developed while an older legal structure was in place. Many of the changes made by the Environmental Protection Act were in fact made necessary by Britain's obligations to implement European Community law addressing acid rain and other air pollution problems. In this section, the pre-1990 legislative position is described.[14]

Those pieces of legislation relevant to air pollution control include: the Alkali and Works Regulation Act 1906; the Clean Air Act 1956; the Clean Air Act 1968; the Control of Pollution Act 1974; the Health and Safety at Work (HSW) Act 1974; and the Public Health Act 1936. Certain provisions of the Electricity Acts, described in Chapter 8, also have some relevance for air-pollution control.

Particularly with regard to complex industrial processes and pollutants other than smoke, the Acts do not specify directly the measures which polluters must take to abate emissions or the standards which they must meet. The statutes are primary legislation which specify general principles of control and devolve the responsibility for activating specific measures to the executive branch of government in the person of the Secretary of State at the DoE.

The formal method for activating these measures is through the use of 'statutory instruments' (secondary legislation). These must be laid before Parliament but are extremely unlikely to be debated, restricting wider political interest in air-pollution control. In practice, years may elapse between the passage of an Act and the activation of the executive branch's statutory powers. Some potential air pollution measures, such as specifying the sulphur content of fuel oil, have never been exercised.

The power to give operational meaning to control philosophies specified in legislation has been vested in bodies such as HM Inspectorate of Pollution (HMIP) which is examined more closely below. HMIP and the local authorities are responsible for the day-to-day implementation of the air-pollution control regime.

THE CONTROL OF MAJOR INDUSTRIAL PROCESSES IN THE UK

Since most of the provisions of the 1906 Alkali &c Works Regulation Act were repealed by a statutory order in 1983,[15] the 1974 HSW Act has been the most important item of legislation relevant to acid emissions. However, the 1906 Act still serves to define emission standards for hydrochloric and sulphuric acids emitted from the alkali processes. For non-alkali processes, Section 5 of the HSW Act states:

It shall be the duty of any person having control of any premises of a class prescribed . . . to use the *best practicable means* for preventing the emission into the atmosphere from the premises of noxious and offensive substances and for rendering harmless and inoffensive such substances as may be so emitted.

Section 5 goes on:

The means to be used . . . includes a reference to the manner in which the plant provided . . . is used and to the supervision of any operation involving the emission of the substances . . .

In the early 1980s, nearly 60 prescribed classes of premises, including electricity works, were defined in Schedule 1 to the Act. Electricity works included industrial generation based on coal or oil where steam output exceeded 200 tonnes per hour. Thirty-six noxious and offensive substances including SO_2, nitrogen oxides (NO_x) and smoke, grit and dust are defined in Schedule 2 of the Act. In 1988, electricity works were replaced by a new category, large combustion plant, to ensure compatibility with the European framework Directive on industrial plant.

The HSW Act provides for the appointment of inspectors, with powers to enter premises at any reasonable time, in order to enforce these requirements. HMIP is the body charged with the responsibility of enforcing Section 5. The HSW Act provides scope for the exercise of a significant degree of administrative discretion by HMIP. The implementation of the Act by the Inspectorate, and the interpretation of BPM, is discussed below.

OTHER POLLUTION SOURCES IN BRITAIN

Emissions other than smoke, grit and dust from non-scheduled works can be controlled only under the statutory nuisance provisions of the Public Health Acts and remain very much a Local Authority task. However, smoke, grit and dust are the main concerns in relation to non-scheduled works. The legislative provisions are found in the 1956 and 1968 Clean Air Acts, supplemented by various regulations made by the Secretary of State.

Vehicle emission regulations remain the primary responsibility of the Department of Transport under the Road Traffic Act and its associated Construction and Use Regulations, although the role of the DoE in policy-making has become more important in recent years largely as a result of the acid-rain debate.

BEST PRACTICABLE MEANS IN BRITAIN

The BPM principle of control dominated the British air-pollution control system for over a century. Under the Environmental Protection Act it has been replaced by the European Community derived 'best available techniques not entailing excessive cost' (BATNEEC) principle. While the use of BPM is, to all intents and purposes, over, the similarities between BPM and BATNEEC and the length of the tradition built around BPM mean that its recent interpretation still provides valuable indicators about the British approach to air pollution control. However, the future evolution of the BATNEEC principle will take place in a European rather than a UK context.

The term 'best practicable means' has rarely, in over a hundred years, been contested in a legal dispute. However, related terms in employment safety legislation, such as 'reasonably practicable', have. Interpretation of these cases leads to the conclusion that 'reasonably practicable' is 'a narrower term than physically possible and involves a weighing of the risk against the measures necessary to eliminate the risk'.[16] Section 34 of the 1956 Clean Air Act defines the term 'practicable' in BPM as meaning:

> reasonably practicable having regard, amongst other things, to local conditions and circumstances, to the financial implications and to the current state of technical knowledge . . .

From this, the following elements of BPM may be identified: (1) the costs of emission abatement are relevant; (2) local circumstances must be taken into account – this might mean the financial implications for a particular emitter or the impacts of emissions in different locations; (3) up-to-date abatement technology should be used, although there is no requirement to push forward the frontiers – 'off-the-shelf' technology only is required; and (4) the risks of pollution damage should be weighed against the costs of abatement. It follows that BPM does not lend itself easily to 'technology forcing' objectives.

In any individual case, the balancing of these different considerations is difficult and the role of the inspecting authorities is crucial. The most recent view of HMIP is that BPM for specific processes:

> will take account of all relevant factors including: the present state of development of relevant control technology; financial implications; current knowledge of the effect of pollutants; local conditions and circumstances where these are relevant in requiring particularly stringent control; and the potential contribution that an individual works may make towards air pollution by a substance that is subject to an air quality standard set in a European Directive.[17]

However, BPM has been an evolving concept and actual practice over recent decades has not necessarily followed the above definition. A more traditional view, expressed in 1963, was that:

> full BPM is reached when the standard of treatment is so high as to result in little or no impact on the community and with no scope for further major improvements . . . the expression BPM takes into account economics in all its financial implications and we interpret this not just in the narrow sense of works dipping into its own pockets, but including the wider effects on the

community ... careful thought has to be given to decisions which could seriously impair competition in the national and international markets.[18]

There was thus a strong sense of balancing costs and benefits and guiding industry towards a condition of low or zero environmental impact, the rate of progress towards that objective being constrained by financial considerations and their impact on competitiveness. Emphasis was also placed on the experience and judgement of inspectors in balancing the individual components of the BPM concept. The approach was also unabashedly technocratic. In 1973, the view of the Chief Inspector was that 'abating air pollution is a technical problem, a matter for scientists and engineers operating in an atmosphere of co-operative officialdom'.[19]

There has been some difference in views on the issue of whether pollution damage needs to be taken into account when determining BPM measures. According to the Chief Inspector in 1981, 'the first duty of the Inspectorate is to prevent emissions to air where it is practicable to do so, irrespective of whether damage is caused'.[20] However, it was also noted that, in practice, the Inspectorate did take account of pollution damage in defining BPM for particular works.

However, the government's view was that 'in setting standards pollution control authorities must take account of the best evidence of the effects of a pollutant'.[21] This symbolizes the philosophical and institutional contradictions which emerged between the Inspectorate and the policy divisions of the DoE during the 1980s. The issue of acid rain and the question of whether, and to what extent, FGD should be fitted at power stations was one of the factors underlying this difference of opinion.

A criticism of the application of the BPM principle in Britain is that is does not create incentives to innovate. This is a characteristic of all methods of control, such as BPM, which focus on technical means, rather than setting emission objectives.

The Inspectorate has been supportive of the technical means approach to regulation, arguing that it gives them more comprehensive, and potentially tighter control over industrial practices. However, specification of technical means inevitably removes a polluter's incentive to innovate and develop cheaper and more effective methods of control. The view of the OECD is that 'the less the regulations concern themselves with technology, the more they facilitate technical change'.[22]

A related issue is that of retrospection – should BPM for new plants be applied to plants which were in existence before current technology was developed? In the past, the view has been that:

in the absence of justified complaints, works which have installed provisional means will be allowed to operate the plants within their economic lives before better measures which have been developed subsequently are demanded,[23]

This principle has no statutory basis, although it might be argued that it is implicit in the need to 'hav(e) regard ... to the financial implications'.

In recent years, there has been evidence of a change in the Inspectorate's thinking about the retrospection principle. BPM notes recently drafted by HMIP now include timetables for bringing standards for existing plant up to the level set for new installations. The deadlines vary between four and eight years depending on the sector involved.

One other strand of thinking has developed since the CEGB proposed retrofitting FGD to some of its largest coal-fired power stations in 1986. Critics had argued that, if FGD represented the BPM at certain power stations, then there was no reason to suppose that it was not also the BPM at other similar ones. The principle of balancing the economic burden of pollution abatement against the benefits has been re-interpreted so that the economic burden is evaluated not for individual plants, but for the appropriate industrial sector as a whole. This also follows the philosophy set out in the European Community's 1984 'Framework' Directive on industrial emissions.

Legisation in Germany

PRINCIPLES OF CONTROL

The German Constitution makes air pollution control a responsibility of Federal authorities in the interest of justice and the welfare of future generations. Environmental protection in general is based on three principles derived from the SPD/FDP Government's environmental programme of 1971. These are:

1. the Vorsorge principle, literally caring beforehand and implying prevention, precaution and anticipation;
2. the Verursacher ('causality') principle, which is difficult to translate directly into English. It is often equated with the polluter-pays principle but its effect is to assign responsibility for effecting a remedy to the legal person causing environmental damage;
3. the Kooperation principle, an expression of German corporatism, requiring government to regulate only after consulting all affected parties.

While originating with the SPD/FDP, these principles have been endorsed by the subsequent CDU/CSU/FDP government which has, if anything, applied the first two principles with even greater diligence.

For the Federal Environment Ministry (Bundesministerium für Umwelt – BMU) Vorsorge has become the dominant principle and was later used to justify officially the measures taken to address the forest die-back problem. However, for environmentalists the same measures may appear to fall short of what 'true' Vorsorge might imply. Vorsorge allows the government both to act before scientific certainty is established and to weaken political opposition. The Verursacher principle is usually considered to be satisfied if the polluting industry rather than the government pays for abatement actions. However, through tax incentives and other types of capital subsidy there is a considerable amount of public expenditure on pollution abatement equipment in the FR Germany.

The three environmental principles are balanced by two more general legal principles. The first is the Gemeinlast (community burden) principle which justifies subsidies and other types of assistance where the polluter is not in a position to pay and abatement measures are required to meet socially defined objectives. Environmentalists would like to increase the application of such measures to small and medium-sized firms. The second is the 'proportionality' principle which requires that the costs of complying with a measure should be commensurate with the benefits derived. This is closely related to the concept of 'economic defensibility' and has now replaced it.

These basic, rather 'soft' philosophical principles have tended to become operationalized in the single requirement on operators of polluting plant to practice abatement with 'state-of-the-art' technology – Stand der Technik.

The interpretation of Stand der Technik, particularly in conjunction with the principle of economic defensibility, has created many headaches for regulators, leading to time consuming negotiations between lawyers, administrators, polluters and, increasingly, the public. The BImSchG defines Stand der Technik as:

the level of development of progressive procedures, installations/equipment or methods of operation, which appear to assure the practical suitability of a measure for the limitation of emissions. In determining the Stand der Technik, comparable procedures, installations/equipment or methods of operation, which have been successfully tried out in practice are especially to be taken into account.

This, and particularly the use of the words 'appears' and 'progressive', has turned out to be a legal nightmare for the administrators required to devise pollution control instruments. The definition is clearly, and intentionally, directed at pushing technology forward, and thus illustrates one of the fundamental differences between the German and UK approaches. In Germany, the tendency is towards technology forcing. Environmental policy thus tends to be technology- rather than science-driven.

The electricity industry has posed regular legal challenges to requirements imposed on them under the TA (Luft) and offered its own definitions of Stand der Technik. The TA (Luft) expands on the BImSchG definition in a way which reassures industry about the prescription of untried technologies. In 1976 the Constitutional Court defined Stand der Technik as 'standards generally accepted if specialists, who have to apply them in practice, are convinced of their soundness'. German courts have therefore made a distinction between Stand der Technik and current best practice technical standards. The Greens and the SPD now wish to introduce the goal of 'beste Stand der Technik' – best state of the art.

Much of the practical interpretation of Stand der Technik has fallen to the more pragmatic German Society of Engineers (Verein Deutscher Ingineure – VDI). For it, Stand der Technik boils down to what industry is able, willing or compelled to pay for pollution control which in turn depends on individual licensing procedures.[24] To encourage as much uniformity as possible throughout the Federal Republic, the VDI has so far issued well over 200 technical guidelines on pollution control, demonstrating its important role in defining the degree and methods of emission abatement achieved in practice.

THE LEGISLATIVE FRAMEWORK

In principle, under the German Constitution, all legislative power remains in the hands of the Länder unless it is superseded by Federal law. While the framework for air-pollution control is increasingly national, much environmental policy, especially with regard to water, coastal discharges, aspects of nature protection and most of planning and agriculture, remains with the Länder. While tensions between Bonn, the Länder and local authorities in environmental matters are chronic, competition over legislative competence has often served to stimulate not only the development of environmental controls, but also the politicization of environmental issues.

The establishment of national standards is both a challenge and an opportunity for the Federal government as it seeks to expand its influence and general management capacity. International agreements, including those reached within the European Community, can assist the Bonn government in that commitments are incorporated directly into domestic law as a Federal as opposed to a Land responsibility.

The Federal Air Quality Protection Act,[25] BImSchG, constitutes the principal legal basis for air-pollution control measures.[26] The BImSchG is enabling legislation which defines the principles upon which pollution control is to be based, allocates responsibilities, enshrines basic obligations and defines decision-making processes in precise terms.

The BImSchG enables the Federal Government to issue both regulations and administrative directives. The different legislative hurdles required to pass a regulation as opposed to an administrative directive have been described in Chapter 6. However, it is important to note that, unlike in the UK, both types of measures must be approved by Parliament. This guarantees a greater party political involvement in environmental debates.

Regulation and directives may specify product standards, ambient air quality standards or emission limits. For all types of measure, there is a considerable emphasis on specifying in detail technical requirements because they are directed at administrators who may not be technical experts. Administrative directives in particular act as guidelines, but must, nevertheless, promote the uniform application of the law. It has been a general objective to move away from locally based, site-specific decisions towards a system of uniform national controls.

The BImSchG prescribes who must be consulted before regulations or administrative directives may be formulated. These include representatives of the sciences, immediately interested parties, the Federal Economics Ministry (BMWi), the highest Länder authorities and, more recently, public interest groups. However, only directly affected parties can challenge BImSchG provisions in the courts.

The principal purpose of the BImSchG is to 'protect man as well as animals, plants and other matter from harmful environmental impacts'. For installations requiring an authorization to operate, it also has the purpose of protecting the same environmental targets from 'dangers, considerable disadvantages and considerable nuisances and to take preventive action against the development of such impacts'. The BImSchG thus provides both for measures taken in defence against tangible threats and for measures taken in anticipation of dangers where causality is not yet established. Terms such as 'considerable' are undefined and are open to interpretation by the authorities and the courts.

The BImSchG also lays down the underlying principle by which its major objectives can be made operational. Paragraph 5 requires that emissions from plants requiring authorization should be reduced using Stand der Technik. Controversy has arisen over the application of Stand der Technik to existing plant and requirements to retrofit pollution controls. Up to 1985, such requirements were allowed only if the associated costs were 'economically defensible'. Since the 1985 revision of the BImSchG, retrospective requirements are permitted where costs are 'proportional' to the expected benefits and where emission reductions are achievable using state-of-the art technology. The 1985 BImSchG revisions also enable the Federal Government to take measures in the general and even global interest.

The BImSchG therefore allows regulators to go beyond what is justified purely by proven environmental damage. However, to do so is not easy and may be subject to challenge in the courts. In general, top-level political support has been required to take anticipatory action.

THE TA (LUFT) – AN ADMINISTRATIVE DIRECTIVE

Administrative directives are the traditional instrument for air-pollution control. They are legally binding only on the officials required to implement them and not on the polluter who is bound by the outcome of the licensing procedure. Directives can be challenged in administrative courts by both the polluter and interested third parties, though not by public interest groups. Directives are often based on, or may even be identical to, technical guidelines produced by the VDI.

The first and most important administrative directive issued under the BImSchG is the TA (Luft). Two other administrative directives, dealing with noise from compressed air hammers and cranes, have been issued.

The TA (Luft) defines air pollutants as 'changes in the natural composition of the air, particularly due to smoke, soot, dust, gases, vapours, aerosols, or odorous substances'. What constitutes 'natural' is not defined and has been a matter of debate in the courts.

The TA (Luft) was initially attached to the GeWO and can be modified relatively quickly to take account of technical developments. Since its first formulation in 1964, it has been modified three times, in 1974, 1983 and 1986. The TA (Luft) has permitted Länder governments to impose more stringent emission standards if they choose to do so. Such standards were, however, applied unevenly throughout the Federal Republic. Operators of existing plants have been able to challenge retrospective requirements over concern about the financial implications using arguments based on the concept of economic defensibility.

Table 9.1   Selected regulations issued under the Federal Air Quality
Protection Act (BImSchG)

Regulation	Name	Remarks
1st BImSchV	Regulation on small furnaces, 1974 Amended 1988	NO_x, SO_2 and CO limits for plant not requiring permits Controls fuel quality
2nd BImSchV	Emissions from dry cleaning plants	
3rd BImSchV	Regulation on sulphur contents in light heating oil	Product standards to reduce emissions from oil and diesel fuel
4th BImSchV	Determination of emissions in polluted areas, 1975	Uniformity of monitoring
5th/6th BImSchV	Emission control officers	
7th BImSchV	Wood dust	
8th BImSchV	Noise from lawn-mowers	
9th BImSchV	Licensing procedures	
10th/12th BImSchV	PCB, PTC and VOC limits, 1977	
11th BImSchV	Definition of the term emission	
13th BImSchV	Regulation on Large Combustion Plant (GFAVo), six parts, 1983	Emission limits for SO_2, NO_x, dust, CO, halogens, carcinogens, applies to new and existing plant over 50MW

During the mid-1970s, the electric utilities began to complain about the interpretation by the different regional authorities of the state-of-the-art criterion specified under the TA (Luft). Licensing procedures were challenged in the courts and retrospective demands to change construction plans were being made at the Länder level. In 1977, the utilities began to press for uniform Federal emission standards, i.e. a regulation, to replace the TA (Luft) in order to protect themselves from leap-frogging regional demands. This was an important mechanism stimulating the development of the GFAVo.[27]

In February 1983, the TA (Luft) was amended to allow the Länder authorities to enforce limits on ambient pollution levels. Incremental emissions of SO_2 or fluorine from any new plant would not be allowed to increase local pollution concentrations by more than 1 per cent.

Outside polluted zones designated as such by the Länder under the BImSchG, concentrations of SO_2 would have to remain below 60 mg/m^3. This was justified with reference to the BImSchG and its requirement to protect plants as well as human health. The building of new plant in designated 'pure air areas' was precluded, making permits to emit in these areas a highly valued possession.

In the 1986 revision of the TA (Luft), tough emission standards were proposed for combustion plant down to a capacity of 1 MW thermal, covering small industry and the service sector. It also specified general waste-disposal

Table 9.2 Emission limit values in the Large Combustion Plant Regulation (GFAVo)[1] for New Plant – milligrammes per normal cubic metre

		Solid fuels	Liquid fuels	Natural gas
SO_2	› 300 MWth	400/85% removal	400/85% removal	35
	100–300 MWth	2000/60% removal	1700/60% removal	35
	50–100 MWth	2000	1700	35
	Fluidised bed	400/75% removal	–	–
	Fuels with variable sulphur	650[3]	650[3]	–
NO_x	In general	800[4]	450[4]	350[4]
	Pulverized fuel/ wet ash	1800[4]	–	–
Particulates		50	50	5

Notes:
(1) The GFAVo also regulates emissions of: halogen compounds; carbon monoxide; and heavy metals. Industrial gases are also covered.
(2) Dry gas, 1013 millibar, 0°C, 3% oxygen for liquid and gaseous firing, 5% for pulverized fuel/ wet ash, 6% for pulverized fuel/dry ash, 7% for grate/fluidized bed.
(3) Plus use of state-of-the-art technology, defined as flue gas desulphurization.
(4) Plus use of state-of-the-art technology, taken to mean selective catalytic reduction for hard coal firing, combustion modification otherwise. In 1985, the Bundesrat decided that plants over 300 MW would be licensed only if they met standards of 200 mg for coal, 150 mg for oil and 100 mg for natural gas.

controls, energy saving and continuous monitoring obligations for the entire stock of industrial plant. This highly controversial legislation, which has by no means been fully implemented, is not discussed here in any detail. Neither the Green Party nor the SPD supported the directive finally issued because it was seen to be too lenient.

REGULATIONS

Table 9.1 shows the regulations which have so far been issued under the BImSchG. Once a regulation is agreed there remains little power of interpretation at the regional level, where authorities function primarily as enforcement agencies.

The Large Combustion Plant Regulation (GFAVo) of June 1983 is a typical Verordnung. It applies to all combustion plant rated over 50 MW thermal. It establishes strict emission limits for seven major pollutants or groups of pollutants. The provisions are shown in Tables 9.2 and 9.3. The use of FGD and catalytic reduction of NO_x are *de facto* requirements at both new and existing power stations given the stringency of the emission standards. The origins and the development of the GFAVo from a political perspective are described more fully in Chapter 10.

Table 9.3   Emission limit values in the Large Combustion Plant Regulation
(GFAVo)[1] for Existing Plant – milligrammes per normal cubic
metre[2]

		Solid fuels	Liquid fuels	Gaseous fuels
SO_2	Residual operating time:			
	>30,000 hours >300 MW	As new plant	As new plant	–
	10–30,000 hours >300 MW or >10,000 hours <300 MW	2500	2500	–
	<10,000 hours	As in site licence	As in site licence	–
NO_x	In general	1000[3]	700[3]	500[3]
	Pulverized fuel/dry ash	1300[3]	–	–
	Pulverized fuel/wet ash	2000[3]	–	–
Particulates		125[4]	50–100[5]	–

Notes:
(1)   The GFAVo also regulates emissions of: halogen compounds (e.g. hydrogen chloride, hydrogen fluoride); carbon monoxide; various metals;
(2)   dry gas, 1013 millibar, 0°C, 3% oxygen for liquid and gaseous firing, 5% for pulverized fuel/ wet ash, 6% for pulverized fuel/dry ash, 7% for grate/fluidized bed;
(3)   plus use of state-of-the-art technology, taken to mean selective catalytic reduction for hard coal firing, combustion modification otherwise. Plants over 300 MW are also to meet the standards described in note 4 to Table 9.2 at the earliest possible date.
(4)   80 mg for lignite;
(5)   depending on volume of flue gas.

While the GFAVo was certainly the most controversial regulation to have been issued, it was merely the 13th in a line of measures enabled by the BImSchG. A small combustion plant regulation was issued as early as 1974.

Implementation

Inspection in Britain

The responsibility for implementing Britain's air-pollution control laws lies with HMIP and the local authorities. Their activities, and the way in which they interpret their responsibilities, are of great importance given the large degree of administrative discretion awarded under the UK system.

The primary focus of this section is HMIP. However the local authorities have been responsible in the past for the control of non-scheduled processes and domestic premises. Following the Environmental Protection Act, they will also be responsible for a second tier of less complex industrial processes specified in the European Community's framework Directive on industrial emissions. Although there has in the past been no officially sponsored mechanism for ensuring co-ordination of the activities and policies of the different local authorities, the Institution of Environmental Health Officers, the professional organization for the responsible local authority officers, fulfils part of this role. In addition, the National Society for Clean Air has many local authority members, providing an additional channel of communication.

HM INSPECTORATE OF POLLUTION

Currently, responsibility for air pollution inspection in England and Wales lies with HMIP which came into being only in April 1987. HMIP is formally a part of the DoE. In Scotland, HM Industrial Pollution Inspectorate (HMIPI) performs the comparable function. Scottish practice is similar to that in England and Wales and is not given separate consideration here.

Prior to April 1987, the responsibilities of the air-pollution control division of HMIP belonged to the Industrial Air Pollution Inspectorate. For simplicity, and to reflect the continuity of philosophy and personnel, the 'Inspectorate' is referred to in this section irrespective of its institutional incarnation. Recent developments at HMIP are discussed in Chapter 13.

Although buffeted by external events in recent years, the Inspectorate has shown a remarkable internal cohesion. From 1863 until the effective demise of the Inspectorate in 1987, Britain had six monarchs, 35 prime ministers – and only eleven Chief Alkali Inspectors. Apart from the first Inspector, all but one of the Chief Inspectors have risen through the ranks. Inspectors are required to have five years of relevant industrial experience prior to appointment. In the past, the Chief Inspector generally operated quite independently of government. Until the 1960s, there was little public attention focused on the Inspectorate, but since then it has come under increasing scrutiny.

A small staff of around 50 is responsible for inspecting 3000 separate processes at 2000 industrial sites. An average of 5 visits per site per year has recently been made, one-third less than a decade before.[28] Although the objective is to visit each site at least once a year, some sites have not been visited at all recently. The falling inspection rate reflects a changing style of inspection, away from routine site visits towards longer and more detailed investigations,[29] as well as a decline in real resources.

As a consequence of the co-operative approach, industry does, on the whole, value the Inspectorate's work. A recent survey showed that industry appears to value the Inspectorate primarily as a source of (free) legal and technical consultancy and for its partnership in dealing with outside bodies – the public, trade unions and local authorities.[30] In addition, it appears to be believed that the Inspectorate will argue industry's case against tighter emission standards coming from the European Community. In future, the Inspectorate's services will no longer come free to industry, as the Environmental Protection Act makes provisions for charging industry for site authorizations.

The Inspectorate's interpretation of BPM for a particular class of works at any time is embodied in a document entitled 'Notes on Best Practicable Means'. The purpose of these notes, developed at the national level, is to form the basis for negotiations between inspectors and the operators of specific works. The notes, which are drawn up in consultation with the appropriate trade associations, can embody the type of equipment to be installed, the way in which it should be operated, the training given to operators,the heights of chimneys, the monitoring of emissions to be carried out and, importantly, guideline emission limits known as 'presumptive limits'. The current interpretations by the Inspectorate is that exceeding a presumptive limit is 'an indication that the best practicable means are not being used'.

However, if a presumptive limit is breached, the law has not necessarily been broken. The only formal ground for an infraction is that BPM have not been used. Thus, breaching a limit occasionally due to operational difficulties or persistent equipment problems which would be expensive to rectify has not, in practice, been regarded as breaching the BPM principle.

Where persistent infractions occur, the Inspectorate may issue an enforcement order under the terms of the HSW Act. This enforcement order may be the subject of appeal to an industrial tribunal. One of the few tests of the BPM principle occurred when British Coal successfully appealed against an enforcement order pertaining to one of its smokeless fuel plants. If an enforcement order is disobeyed, the Inspectorate may order the closure of the process concerned or may prosecute.

However, the Inspectorate is generally reluctant to initiate prosecutions unless operators are persistently or wilfully negligent. It has become virtually an article of faith that persuasion and education, rather than coercion, are the most effective ways to implement pollution controls. This attitude is reinforced by the fact that all inspectors must have industrial experience, and the way in which long careers inside a self-contained body are the norm. One Chief Inspector wrote in 1963 that:

> BPM are comprehensive provisions which, like some systems of Contract Bridge, require a deep understanding and lots of experience with co-operative partners.[31]

The rate of recorded infractions is now running at 20–30 a year, compared with a peak of about 60 per year during the late 1970s. One or two prosecutions per year are typical, though there were as many as seven in 1977.

The Inspectorate's policy of using prosecution only as a last resort has effectively pre-empted any role which the courts might play in influencing the application of Britain's air pollution control laws. The few prosecutions which are instigated are usually, from the Inspectorate's point of view, conducted on safe ground and defendants commonly plead guilty (receiving small fines) rather than contesting the charge.

THE INSPECTORATE AND THE PUBLIC

In the 1960s, as prosperity grew and air pollution began to be regarded as something other the symbol of people at work, the Inspectorate began to get a reputation (from those who had heard of it) for remoteness, insensitivity and connivance with industry. This led to conflicts with local authorities and pressure groups and exposure in the media. Outspoken criticisms of the

Inspectorate and its methods were made both inside and outside Parliament in the late 1960s and early 1970s.[32] The Royal Commission on Environmental Pollution was, however, supportive of the Inspectorate and the scale of the controversy provoked by the Inspectorate diminished in the late 1970s. Criticisms made during the 1980s have focused more on falling inspection rates, lack of resources, low morale and recruitment problems rather than on the style of operation.

POLITICAL PRESSURES AND THE INSPECTORATE

The Inspectorate's reluctance to take forceful measures against industry stems not only from its own internal traditions, but also from outside pressures. The political climate within which the Inspectorate operates has on occasions affected its ability to enforce its own definitions of BPM.

The failure of the Inspectorate to secure an enforcement order against British Coal's South Wales smokeless fuel plant has already been mentioned. British Coal based its defence against the enforcement order on the inflexibility of working practices negotiated with the National Union of Mineworkers (NUM), an argument well adapted to impressing the local industrial tribunal.

Failures to prevent the electricity supply industry from breaching presumptive dust emission limits can be attributed to the political nature of power station fuel supply. For instance, when a major dual-fired power station was converted from gas to coal-firing in the late 1970s, it severely breached emission limits because the condition of its dust arrestment equipment had deteriorated. The Inspectorate's order to revert to gas-firing was countermanded at a political level because of the energy policy commitment to increase coal burn at the expense of oil and gas. Similar incidents have subsequently taken place.

These examples symbolize the Inspectorate's subordinate role within the wider policy process. When the CEGB proposed the retrofitting of FGD to 6000 MW of power-station capacity in 1986, the Inspectorate was not consulted. That the Inspectorate was pre-empted over the question of requiring FGD on power stations by the CEGB, a body which it was formally charged with regulating, underlines its lack of assertiveness in politically charged areas. In the very early 1980s, Chief Inspectors had hinted that FGD might be a requirement for new power stations, but did not venture a forthright statement to that effect.

Although HMIP is now formally a part of the DoE, it enjoys an ambiguous relationship with the Department's policy divisions. Mainstream civil servants are sceptical of the Inspectorate's capability to operate in a more political environment, and HMIP staff have played no direct role in European Community negotiations. This lack of technical input partly explains the UK's scientific rather than technological policy posture. Equally, air inspectors are resentful of perceived loss of status within HMIP. Patchy information flows with DoE's mainstream divisions have not helped the policy-making process.

Implementation in Germany

Enforcement of legislation has remained much more fragmented and decentralized in Germany and there has never been a national inspectorate as in the UK. Enforcement is often identified as the weakest point in an otherwise ambitious system. This has been of critical importance in determining the course of environmental politics in West Germany.

The authorization of plant licences is the perogative of individual Länder, which have had scope to exercise a considerable degree of interpretation of such Federal legislation as is implemented by administrative directive. There is a great regional diversity in the precise methods of implementation and enforcement. In some Länder, technical experts are empowered to carry out regulatory duties under political supervision: in others, regulation is an administrative responsibility with technical experts playing an advisory role.

Regulators are generally assisted by non-governmental bodies such as the prestigious VDI and the regionally based Technical Inspection Offices (Technische Überwachungsvereine – TUVs), both of which advise industry as well as government and are thus major actors in the environmental control system.

Compliance with environmental standards is enforced either by the authorities themselves through occasional visits and reporting procedures, or by inspection through VDI or the TUVs. For example, the inspection of boilers for safety purposes began in 1859 and was carried out by the TUVs. They still play a major role today in inspecting industrial plant and testing vehicle emissions. Local chimney sweeps remain responsible for small emitters. The growing use of continuous emission monitoring is beginning to change the nature of inspection.

THE IMPLEMENTATION GAP

Given the structure of this system, the potential for the development of an 'implementation gap' is obvious.

As the green movement grew during the 1970s, one of its most important accusations against industry and government was that of slow and inadequate enforcement of environmental measures. This was explained with reference to the legal and institutional complexity of the environmental protection system, the gap separating legislators from implementation agencies within the Federal structure and the bias of the legal system towards industry and pollution. Even German political culture with its emphasis on formal law and reliance on state action was cited as one of the underlying causes.[33] An analysis of the air pollution control system in the early 1980s concluded that:

> although the FR Germany has the most restrictive and severe legal regulations governing air pollution control in Europe, as well as the most developed administrative system devoted to its implementation and has put into effect the most comprehensive licensing procedures for large polluters, there has been no progress in air pollution abatement since 1974.[34]

This began to change after 1982. The demand for a more Federally-based system of regulation has therefore sprung, in part, from this situation and thus from within the air pollution control regime itself.

Conclusions

The comparison between the air-pollution control systems in Britain and Germany reveals a number of important differences which directly influenced the outcome of the acid-rain debate. The German regime, which makes clearer distinctions between policy planning, policy-making and implementation, has different strengths and weaknesses from that in the UK.

The existence of a recent written constitution in Germany ensured the clear articulation of the philosophical basis for environmental policy. This lays down how pollution control must be justified and what broad advice must be sought. Major regulatory changes are not possible without Parliamentary approval. With no written constitution, the British system of control remained pragmatic and reactive. There is comparatively little political interest in air pollution control since important provisions may be altered at an administrative level. Even where primary legislation allows the Government itself to specify emission controls, this can be done by laying orders before Parliament without any debate taking place.

Given political pressure in Germany, the Vorsorge principle and its interpretation by the courts allows technology forcing through the prescription of emission standards, weakening the need to establish firm causal links between emissions and environmental damage. However, in Britain, scientific evidence for environmental damage remains the major official criterion for action.

The German air-pollution control system is considerably more legalistic, with requirements specified in great detail for a limited number of largely technological objectives. These are derived from very broad and ambitious principles. The British system of control is, by comparison, vaguer and less ambitious in its legal foundation and allocates more discretionary powers to appointed regulators who perform their tasks with relatively little interference from outside. This difference is fundamental and cannot be altered without major political effort.

The courts are a significant player only in Germany, where their use by interested third parties during licensing procedures for industrial facilities helped to promote tighter standards of control and regulatory measures themselves can be challenged if they are believed to be unconstitutional.

More importantly, once the environment had become an active political issue in Germany, policy development was encouraged by competition for jurisdiction between the Federal and Land authorities. Concurrent legal competence for pollution control also promoted the adoption of uniform Federal standards as Bonn attempted to extend its powers and industry has sought to escape from what it believed to be arbitrary, locally imposed demands for increasing abatement.

A centralized national regulator in Britain could, in principle, have acted more ambitiously, but had neither the incentives nor the resources to do so.

Potential activism within the regulatory regime was severely constrained by political and economic factors. In Britain, the lack of high level political interest in air pollution control issues after the 1950s meant that the regulatory authorities were isolated and unassertive, demonstrating that under-resourced regulatory regimes without significant political and legal support are unlikely to be effective innovators.

Placing German pollution control in a wider context, a decentralized regulatory regime tied to more precise and complex legal instructions was stirred into action by a well-resourced interdisciplinary bureaucracy. This helped to maintain momentum in environmental policy even when political interest declined. This interest was rekindled in the early 1980s by energy/environmental policy problems which air-pollution control promised to address. In Britain, the potential of the national regulatory authority remained frustrated by a lack of interest from both administrators and politicians and, in the 1980s, by outright opposition from the electricity supply industry.

Notes

1. Much of this section is derived from: E. Ashby and M. Anderson (1981), *The Politics of Clean Air*, Monographs on Science, Technology and Society, Clarendon Press, Oxford.
2. E. Ashby and M. Anderson, op. cit.
3. *Report of the Committee on Air Pollution* (1954), Cmnd 9322, HMSO, London.
4. A. Auliciems and I. Burton (1973), Trends in smoke concentrations before and after the Clean Air Act of 1956, in *Atmospheric Environment*, Vol. 7, pp. 1063–70.
5. Royal Commission on Environmental Pollution (1976), Fifth report, *Air Pollution Control: An Integrated Approach*, Cmnd 6371, HMSO, London.
6. Reported in Annexe 18 of *Inspecting Industry: Pollution and Safety, Efficiency Scrutiny Report* (1986), HMSO, London.
7. Department of the Environment (1982), *Air Pollution Control*, Pollution Paper No. 18, HMSO, London.
8. Health and Safety Executive (1982), *Industrial Air Pollution: Health and Safety 1981*, HMSO, London.
9. This section is based on: K-G. Wey (1982), *Umwelt Politik in Deutschland*, Westdeutscher Verlag, Opladen; F. Spielberg (1984), *Reinhaltung der Luft im Wandel der Zeit*, VDI Verlag; G. Olschovy (1978), *Natur und Umweltschutz in der BRD*, Parey, Berlin; and S. Boehmer-Christiansen (1989), *The Politics of Energy and Acid Rain in FR Germany*, SPRU Occasional Paper No 29, Brighton.
10. M. Iljra (1981), Luftverunreinigung und Immissionsschutz in Frankreich und Preussen zur Zeit der frühen Industrialisierung, in *Technikergeschichte*, Band 48, VDI.
11. E. Müller (1986), *Die Innenwelt der Umweltpolitik*, Opladen.
12. K-G. Wey (1982), *Umwelt Politik in Deutschalnd*, Westdeutscher Verlag, Opladen.
13. R. Mayntz in Mayntz and F. Scharpf (eds), 1975, *Policy-making in the Federal German Bureaucracy*, Elsevier.
14. For a full description of legislative provisions in Britain, see J. McLoughlin and M.J. Forster (1982), *The Law and Practice Relating to Pollution Control in Great Britain*, Graham & Trotman, London.
15. *Health and Safety (Emissions into the Atmosphere) Regulations 1983*, SI 1983/943.
16. *Halsbury's Statutes* (1986), Vol. 19, 4th ed, Butterworths, London, p. 617.

17. HM Inspectorate of Pollution (1988), *Best Practicable Means: General Principles and Practice*, BPM 1, Department of the Environment, London.
18. *100th Annual Report of the Alkali Inspectorate for 1963* (1964), HMSO, London.
19. *110th Annual Report of the Alkali and Clean Air Inspectorate for 1973*, (1974), HMSO, London.
20. Health and Safety Executive (1982), *Health and Safety: Industrial Air Pollution 1981*, para. 132, HMSO, London, p. 14.
21. Department of the Environment (1984), *Controlling Pollution: Principles and Prospects*, Pollution Paper No. 22, HMSO, London.
22. OECD (1985), *Environmental Policy and Technical Change*, OECD, Paris.
23. *100th Annual Report of the Alkali Inspectorate for 1963* (1964), HMSO, London.
24. P. Davids (1984), *Die Grossfeuerungsanlagenverordnung*, VDI Technischer Kommentar.
25. The full title is the 'Law for protection against harmful environmental impacts from air pollution, noise, vibration and similar processes'.
26. J. Salzwedel and W. Preusker (1982), *The Law and Practice Relating to Pollution Control in the FR Germany*, Graham & Trotman.
27. P. Davids (1984), *Die Grossfeuerungsanlagenverordnung*, VDI Technischer Kommentar.
28. Health and Safety Executive (annual), *Health and Safety: Industrial Air Pollution*, HMSO, London.
29. *ENDS report 124* (May 1985), p. 14.
30. *Inspecting Industry: Pollution and Safety – Efficiency Scrutiny Report* (1986), Annexe 15, HMSO, London.
31. *100th Annual Report of the Alkali Inspectorate for 1963* (1964), HMSO, London.
32. These are epitomized by a stinging critique made by the journal Social Audit in 1974. See M. Frankel (1974), *The Alkali Inspectorate: The Control of Industrial Air Pollution*, Social Audit, London.
33. K-G. Wey (1982), *Umwelt Politik in Deutschland*, Westdeutscher Verlag, Opladen.
34. Weidner and Knoepfel, cited in Bolsche (1984), *Das Gelbe Gift*, Rohwolt, Hamburg, p. 298.

PART III
The story of acid rain in Germany and Britain

10 Forests and power: German precaution

Introduction

This and the following chapter tell the stories of the acid-rain debates in Britain and Germany, illustrating both the origins of concern and the measures adopted to reduce acid emissions. The two chapters draw on the comparative themes developed in Chapters 4 to 9.

The West German acid-rain story, which covers the period 1977–89, emphasizes the salience of clean-air policy in general, as well as the considerable amount of official attention devoted to abating acid emissions after 1981. This was a time when West Germany was experiencing serious political difficulties, as well as growing public concern about the dying of its forests, allegedly as a result of air pollution.[1]

The centrepiece of the German story is the adoption of a comprehensive clean-air package in 1982–3.[2] The main ingredient, and the first legal instrument to be fully implemented, was the Large Combustion Plant Regulation (Grossfeuerungsanlagenverordnung-GFAVo). The first signs that the GFAVo was on the way came in July 1982 when an international policy reversal was announced by Foreign Minister Hans Dietrich Genscher in Stockholm. Shortly afterwards, the Federal Cabinet accepted the GFAVo, in principle, into its legislative programme and in the summer of 1983 it became enforceable law.[3] Tables 9.2 and 9.3 show some of the GFAVo requirements.

Originally drafted in a much milder form in 1978 as part of a Federal Energy Programme, the GFAVo subsequently became the model for the EC Large Combustion Plant (LCP) Directive, described more fully in Chapter 12.

The electric utilities were therefore faced in 1983 with uniform, nationwide emission limits which were among the toughest in the world, as well as detailed monitoring rules requiring expensive electronic equipment. The implementation of these regulations was made subject to strict timetables and failure to do so was sanctioned by severe penalties. The new limits were nevertheless adjusted to specific fuel types and not without some temporary concessions to coal. There can be no doubt that the GFAVo deserves to be called a 'technology-forcing' regulation.

The clean-air package also dealt with motor vehicle emissions. By 1982, these were already a European matter, although at a very permissive level with the EC adopting regulations negotiated with industry within the UN Economic Commission for Europe (UNECE) in the interest of free trade. These regulations had already made European emission regulations considerably less onerous than those in the USA and Japan. These technical norms could be tightened in a legally binding form only through additional European Community (EC) legislation and thus had to be negotiated with all EC partners in spite of early German threats to go it alone. Such an agreement had not yet been fully finalized in early 1990, although political agreement on tough standards requiring three-way catalytic converters for all new vehicles after 1992 was finally reached in late 1989. (A fuller story of these negotiations is told elsewhere.[4])

In the mean time, the FR Germany tried to implement its own preferences for environment-friendly cars on the basis of a voluntary regime supported by fiscal incentives. For mobile sources, therefore, Germany did not, in the end, attempt legal unilateralism. Instead, the preferred technical solution had to be painfully negotiated in Brussels, where, as for power stations, determined British opposition had to be overcome.

In constrast, the GFAVo constituted both a deed done unilaterally and a model to be followed by others. It was confidently expected in 1983 that European partners would be required to follow the German example in good time.

The chapter divides into three sections. The first outlines the growing links between energy and environmental policy before 1982 when the 'integration' of energy and environmental considerations by means of 'Umweltpolitik'[5] was encouraged by administrative developments, while the political context remained more hostile.

The second section deals with the formation of an alliance in favour of clean coal in response to both Waldsterben (forest dieback) and changing political priorities. This allowed government to legislate in support of a highly ambitions clean air programme which was almost fully in place by mid-1983. The adopted measures not only served to improve air quality but also facilitated, or at least did not contradict, several other public policy goals, including important aspects of German energy policy. In addition, acid-rain abatement policy, as the major single item on the environmental agenda of the new government coming to power in 1982, helped to maintain the functioning of a political system which had been showing signs of instability.

The last section deals with implementation after 1983 and briefly discusses the role played by economic incentives in implementation, as well as the European dimension in the German debate.

The foundations of German clean air policy 1977–82

The legacy of the 1970s

Tall chimneys were adopted in Germany as well as other countries to ensure the 'harmless' dispersal of power-station emissions. In this way, major emitters were able to comply with air quality standards which were being defined in Germany during the mid-1970s, first at the Land level and subsequently by the

EC. However, the siting of new plant was becoming increasingly more difficult. Pressure on polluters to reduce their emissions, both of acid gases and radioactivity, continued to grow throughout the 1970s and necessarily expressed itself most strongly at the local level where it could be dealt with by the Länder under either the TA (Luft) or the Atom Act.

In marked contrast to Britain, the environmental policy-making capabilities of government had been growing stronger throughout the 1970s. The history of official concern over air pollution and damage to plants and forests can be traced back to the late 1960s when the Federal Ministry of Health (BMG), then in charge of public hygiene and thus air pollution, stated that 'sulphur dioxide ... can lead to breathing difficulties, cause grave damage to plants, especially endangering forests'.[6]

Energy policy and the emerging green movement

Following the first oil crisis, it became official policy to reduce the amount of oil used and to develop an ambitious nuclear programme. The Federal Economics Ministry (BMWi) at that time agreed to prescribe flue gas desulphurization (FGD) as the 'state-of-the-art' technology for large oil-fired power stations. Government failed, however, to persuade the coal burning utilities to follow suit, even with the promise of grants on offer for GDF units. Here, as well as with the powerful car industry, trade-offs between the interests of industry and those of the political system had yet to be discovered and negotiated.

While there had been some progress in Federal air-pollution regulation for air-quality standards during the 1970s, (see Chapter 9) more stringent emission standards did not emerge until political pressures changed perceptions and interests. Official attention to clean air declined during the latter half of the 1970s partly because the attention of the Federal Interior Ministry (BMI), the Ministry responsible, was focused on the growing difficulties with the social acceptance of the nuclear programme.

An implementation gap for environmental policy between that which had been promised in the early 1970s and that which was being achieved, had developed. This attracted the attention of citizen action groups and the green movement (see Chapter 5), which saw coal-fired power stations as the next target after nuclear reactors. It is most unlikely, however, that the environmentalists alone would have been able to reverse policy.

In contrast to Britain, the action groups were able to make use of the administrative courts in their challenge to regional emissions regulation. There, according to an inside observer, 'the students of 1968 had meanwhile become the judges'.[7] They thus initiated a cycle of ever-tightening emission standards fixed at the Land level. The response of the utilities was to call for Federal regulation.

A Federal regulation covering SO_2 emissions from power stations promised redress and an undertaking to develop such an instrument made its way into the BMWi's Third Energy Programme of 1978. The utilities' request was for Federal emission standards unchallengeable in the courts, to replace the TA (Luft). The original purpose of the 1978 draft therefore was to remove the influence of these courts and to weaken the impact of the environmental lobby at the grass-root level. The same motivations underlay the GFAVo of 1983, although a broader set of political mechanisms was by then in operation.

Agreement on the level of Federal standards could not be reached until 1983. In the late 1970s, the BMI failed to have its proposals, which covered only new plant, accepted in Cabinet because economic considerations still outweighed environmental ones. Environmental controls were still perceived as an obstacle rather than a stimulus to economic growth. However, this perception was beginning to change.[8].

Just as economic problems were weakening the environmental cause, international developments came to the aid of the environmental managers in Bonn. In response to renewed Swedish pressures, the Federal Environment Office (UBA) held, in 1978, a hearing on forest damage and the long-range transport of SO_2 in preparation for negotiations on the UNECE Convention on Long Range Transboundary Air Pollution (see Chapter 2). This raised the possibility of substantial domestic environmental damage arising, not from the future failure of reactors or discharged radionuclides, but from subsidized coal-fired power stations.

While the officials dealing with radioactive emissions in the BMI had been kept busy defining tougher safety requirements during the late 1970s, those dealing with acid pollutants had spent their time squabbling over the revision of relevant legislation and the technical severity of standards. One unmistakable input in this debate was an awareness that state-of-the-art technology, already in use abroad and especially in Japan, was achieving emission reductions greater than those of German industry. This aroused a competitive streak which was not tempered, as it would have been in the UK, by any cultural or legal resistance to the concept of technology forcing.

Aware of these internal controversies, the Interior Committee of the Bundestag held a public hearing on air-pollution control in January 1980 in which draft revisions to both the BImSchG and the TA (Luft) were debated. During this hearing, all parties to the environmental debate produced their own experts. Parliamentarians heard a range of views which described the drafts as either too lenient or too severe, as technologically feasible or altogether impracticable. The administration clearly needed guidance from the politicians and the electorate on how to proceed.

Politics in the early 1980s

The 1981 elections resulted in a shaky SPD/FDP coalition remaining in office. Franz-Josef Strauss, the leader of the Bavarian-based CSU failed in his bid to become Chancellor. Uncertainty about energy policy continued and decisions on acid rain were postponed until a convergence of nuclear, environmental, political and economic policy objectives became possible. Any concerted policies to reduce acid emissions would have to be compatible with a range of other issues close to the heart of government. This proved to be possible only after a major shift in political power. By 1981, the Green Party had entered regional politics with nuclear power as its primary target (see Chapter 5), profoundly worrying both the political establishment and the utilities.

External pressure from the Scandinavians had so far proved too weak to stimulate policies which would significantly mitigate acid emissions. The idea of serious domestic damage had first to be officially recognized and communicated widely. This was achieved above all by two experienced and skilful political tacticians, Hans-Dietrich Genscher former head of the BMI during the heyday of environmental policy planning, and by then in charge of foreign policy, and the late Franz-Josef Strauss, Prime Minister of Bavaria, Chairman of the Bundesrat and a former Nuclear Minister. As early as 1980, Bavaria had begun to protest in the Bundesrat against acidification caused by long-range transport of air pollution.

Waldsterben and politics 1982–83

Forest death and scientific uncertainty

A third of the area of the FR Germany is covered by forest. Forests are a valuable economic resource which are exploited for timnber, grazing and hunting and which attract many other leisure activities. Problems with the forests, but not their causes, had been recognized for some time, but the phenomenon became widely associated with air pollution only in 1981–2. Trees were showing visible signs of ill health: their needles were turning yellow and dropping in excessive numbers. Inspired by scientific hypotheses which began to gain a wide currency, politicians, the media and environmental activists began to conclude that the problems of the forests, as well as the observed acidification of soils and water, were directly attributable to the long-distance transport of SO_2. (See chapter 3)

In November 1981, the influential news magazine, *Der Spiegel*, published a cover story showing sickly conifers in front of smoking chimneys, declaring that 'Der Wald stirbt' (the Forest is dying).[9] It predicted, on the basis of a hypothesis proposed by soil scientist Bernhard Ulrich and presented in the media as an established fact, that large tracts of German forests would be dead within five years. The uproar which followed produced vocal and often very emotional demands for immediate action, especially from forest farmers and powerful hunting associations in Southern Germany, representing people not known for their support for the radical Greens. Public perceptions of forest damage were greatly influenced by media presentation. Questionnaires undertaken at the time showed that people tended to believe that forests which they had not themselves seen, were worse affected than those with which they were familiar.

Alerted to the signs of forest damage in 1981, the Federal Agriculture and Forestry Ministry (BMELF) began to collect detailed data on forest health in June 1982. It asked the managers of large private and state owned forests, covering about 60 per cent of the total, for damage estimates. While reluctant at first to believe that there was much damage, forestry officials soon noticed a great deal. The 1982 forest survey used only three categories of damage intensity defined by the percentage of forest area covered with affected trees. It concluded that 75 per cent of forested area was damaged slightly, 19 per cent moderately and 6 per cent seriously.

All the damage observed in the early forest surveys was attributed to air pollution. The government and politicians ignored alternative hypotheses. Until 1985, statements tended to protect official policy from hypotheses in which air pollution played no role or was only a minor causal factor.[10]

Annual forest damage maps based on a more sophisticated assessment of damage have been published annually since 1983. Forested areas are now divided into five categories on the basis of aerial surveys in which trained observers are asked to judge tree characteristics. Errors in the earliest mapping exercise are now officially admitted, and explanations for the problem are phrased much more carefully. In 1988, 1 per cent of forested area was classified as seriously damaged, 14 per cent as moderately damaged, 37 per cent as slightly damaged and 48 per cent as undamaged. The total proportion of forests either moderately or seriously damaged peaked at 19 per cent in 1985, double the 1983 level. Since 1985, the area of forests falling into the two highest damage categories has apparently fallen from 19 per cent to 15 per cent.

The hypothesis that SO_2 was the direct cause of forest damage, only true for severe spruce damage, aroused much anxiety. Probably because it proved so politically useful, the idea became an accepted truth almost overnight and was generalized to cover the entire Federal Republic and all tree species. Scientists, such as those advising the BDI, who pointed out other possible explanations were not believed at the time. German scientists knew that the early forest surveys in particular lacked scientific validity as they simply listed certain poorly understood effects in a rather imprecise manner.

Politicians from North Rhine-Westphalia (NRW) who as much as mentioned other hypotheses, such as drought or high ozone levels, were accused in the press of defending vested coal interests. In early 1982, there was an outcry of disbelief in Bavaria, when a NRW Minister dared to publicly challenge the SO_2 explanation by blaming vehicle emissions. On the other hand, the German scientific community did not come out in opposition to the major SO_2 control measures adopted. Like the population at large, German scientists believed that precaution was a wise policy to adopt.

From the beginning, the cause of forest damage had been a matter of debate among the scientific community in Germany as much as in Britain. In both countries, science benefited from the consequent flow of research funds and public attention. The difference was that, in Germany, the government believed that the evidence justified preventive action.

Since the Southern half of the FR Germany is much more forested than the North and North West, attention paid to forestry at the national level, i.e. by the BMELF, meant more attention and funds for the southern Länder, irrespective of whether forests are owned privately or by the Land governments. Franz-Josef Strauss's Presidency of the Bundesrat and the growing strength of the Bavarian-based CSU in the German political system in 1981–2 help to explain the growing interest of Southern Germany in the forest dieback issue.

One reason for the CDU/CSU's growing interest was the opportunity to apportion blame for forest damage to the existing coalition Government. The policy of tall chimney stacks to disperse emissions was seen as partly responsible, and an accusing finger was pointed at NRW and the Saarland, West Germany's main coal-mining regions and traditional SPD fiefdoms. Dismay was particularly strong in the South, because it did not produce much SO_2. Enormous pressure was put on politicians and industrialists, including motor manufacturers, to accept the responsibility for forest damage and to act quickly in order to save a major national resource.

The threatened forests were both large plantations covering whole mountain ridges, and small plots scattered among fields and pastures. Most were less than 100 years old and had been established, often by statute, after the virtual deforestation of earlier centuries. Forest plantations today grow on the poorest soils, especially on steeper slopes, as well as near major cities where they have been planted quite recently to provide 'green lungs' for densely populated regions. In the mountains, overgrazing by deer and cattle roaming 'natural' pastures presents genuine and more recently acknowledged problems. By assuming that forest damage was entirely due to air pollution a very credible cost-benefit balance in favour of very stringent abatement could quickly be constructed by economists.[11]

As more carefully weighed evidence moved blame away from SO_2 towards NO_x and ozone, one of Southern Germany's major industries, motor vehicle manufacture (companies such as BMW, Porsche and Daimler Benz), also had to defend its image and search for viable technical solutions. This solution existed already in the USA and Japan in the form of catalytic converters and merely required adoption on a European-wide basis.

During late 1981 and early 1982, therefore, a strong demand for tougher emission controls for stationary as well as mobile sources built up both from the wider public and from an important group of political interests. The Southern Länder especially worked hard for the widespread acceptance of forest damage as a powerful threat image. While earlier CDU/CSU support for anti-pollution measures may have been tactical rather than genuine, this was by then no longer the case. The electorate and the political parties had 'greened' and they demanded action.

Growing political instability in 1982

The wider political context significantly helped to influence the course of the Waldsterben debate. In 1981, it became clear that new American Pershing missiles, unwanted by a large proportion of the electorate, were to be accepted on German soil. Important nuclear decisions were outstanding and new reactor construction had to be made palatable.

The missile crisis encouraged the Federal government to direct attention towards the issue of forest damage and clean air. The Pershings were greeted with such a degree of opposition that the FR Germany's two largest non-mainstream political groupings, the green and the peace movements, came close to joining and were widely perceived as threatening political stability. The coalition between the SPD and FDP was showing open signs of strain. A protective, unifying and popular policy theme, such as action to prevent Waldsterben, was becoming attractive to the Bonn government.

Grass roots opposition to nuclear power was still growing, even inside the SPD, and 'Endzeitstimmung', an apocalyptic feeling of doom, prevailed. People were disillusioned with politics and government. Unemployment was rising, the Deutschmark was losing value and inflation began to disturb a society especially sensitive to this issue.

The first step towards reversing official German environmental policy was taken by Hans-Dietrich Genscher in June 1982, when he committed the FR Germany to the '30% Club' for SO_2 abatement. His party, the FDP, had strong incentives to promote tough measures to curb atmospheric pollution. The growing electoral challenge of the Greens, the FDP's most important competitor at the ballot box at that time, alone must have been an important consideration. In essence, therefore, the FDP and not the Green Party was the prime mover behind the GFAVo in 1980–2. As will be described, the Bavarian CSU was primarily responsible for its eventual stringency and implementation.

A change in policy at the top had become possible because with Waldsterben causing so much alarm, BMELF could no longer stay aloof. Instead, very much in contrast to the role played by the Forestry Commission in Britain, it became the prime mover at this point in the decision-making process.[12] In addition, a strong position in favour of tough standards had already been adopted by the Bundesrat. Here the conservative Länder had a majority of votes and were clamouring for action on several energy related environmental issues.

The BMI's role at this point was also important because it faced two important sets of energy-environment decisions. The construction of the first of the FR Germany's four convoy nuclear reactors, Isar 2 in Bavaria, was due to start in August 1982, but the required licence had still not been provided by July because the FDP Minister was deeply concerned about safety questions. He agreed to license the plant in July 1982 only after substantial pressure to go ahead had been applied by ministries such as BMWi and the Federal Ministry for Research and Technology (BMFT) and after a further tightening of safety standards. Political movement on the Waldsterben issue surely offered a chance to deflect attention away from this controversial decision.

On the air pollution front, therefore, the SPD/FDP Cabinet accepted a relatively lax version of the GFAVo into its legislative programme at a meeting chaired by Chancellor Helmut Schmidt in September 1982. This decision was based on advice from four FDP-controlled ministries: Interior; Economics; Agriculture and Forestry; and Foreign Affairs. However, the draft legislation applied only to new plants and not to existing ones, significantly softening the likely impacts on coal markets. Plants still under construction but already possessing a permit were also exempt under this proposal.

Given the public outcry about forest damage, the pressures from the FDP and discontent about environmental policy coming from within his own party Chancellor Schmidt's change of mind is not surprising. The very existence of Waldsterben as a national issue distracted public attention from intensifying anti-rearmament protests and foreign policy problems.

However, those who had hoped in mid-1982 that the GFAVo would not have 'real teeth', were to be disappointed. The dispute between the various Ministries involved had by no means been fully resolved and the precise contents of the GFAVo remained a subject for negotiation until Spring 1983, when the balance was finally tipped towards stringent emission standards for all plants above 50 MW, the inclusion of brown coal and a precise timetable for retrofitting or closure of existing plants.

The fall of the government

In October 1982, the government of Chancellor Schmidt collapsed following a tactical vote of no-confidence when the FDP decided to switch its support to the CDU/CSU. Gerhart Baum, the FDP Interior Minister from the left of his party, resigned immediately, thus making room for a newcomer in the BMI which was responsible for protecting the environment.

Just prior to this collapse, however, the SPD/FDP coalition had adopted a very progressive environment protection programme which contained far more than just the draft GFAVo. The programme included most of the important elements of the new government's subsequent environmental policy:[13]

1   the BMI was required to begin an initiative at the intergovernmental level (i.e. through the OECD, UNECE and EC) to make German emission and product standards the guidelines for international environmental policy;
2   within the wider context of international transboundary pollution, national efforts were to be made which would act as an example for others to follow;
3   the TA (Luft) was to be amended with the objective of improving the protection of human health as well as extending protection to sensitive plants and animals. Existing pollution sources in heavily populated areas were also to be cleaned up;
4   a GFAVo was to be adopted with the aim of introducing new, very strict emission limit values for SO_2, NO_x and particulates covering plants using gas, oil and coal. Existing plants should be either retrofitted or shut down within ten years; and
5   there was to be an effort to get vehicle emission standards tightened at the European level.

The above programme differed little from that which the Greens had proposed for some time and which the new government in fact implemented. In the environmental policy realm, this illustrates what in German is referred to as 'Themenklau', the theft of one party's policy themes by another.

It has been argued in Germany that one of the main reasons underlying the fall of the SPD was its underestimation of the strength of the ecological movement. Its final attempt to bring green issues back on to its agenda came too late according to this argument. The fatal reluctance of the SPD to change course earlier very much reflects the influence of German industry.

The responses of industry

In 1982, the obligations of stringent emission standards for lignite as well as for hard coal, and for existing plants as well as new ones, were yet to come. These were put in place only in 1983 by the new, strongly pro-nuclear government in Bonn which simultaneously helped to orchestrate the public outcry about forest damage.

The position taken by the major utilities, concerned about the costs of their energy sources and mostly reliant on both coal and nuclear power, became critical during the final stages of the Waldsterben debate. The proposals for the GFAVo were initially perceived as an attack on hard coal and lignite, and as such on domestic fuels and fuel supply security. This link had direct party-political implications because the nuclear industry and its political supporters tended to equate the Social Democrats with the coal lobby.

The forest damage issue was not clear-cut from the utilities' point of view. On the one hand, they still had to construct the recently approved convoy of nuclear plants against a background of public hostility. In this context, the widespread public perception that coal-fired power stations were dirty and environmentally damaging would help to direct attention away from the nuclear programme and create a climate within which nuclear power might be more readily accepted. On the other hand, the costs of implementing an acid emissions clean-up programme, which seemed to be the inevitable consequence of public concern about forest dieback, would be extraordinarily expensive.

In the event, the coal-burning electric utilities decided to fight the tightening of the draft GFAVo requirements as adopted by the incoming CDU/CSU/ FDP Coalition. In response, the argument that nuclear power would help to reduce acid emissions was used and played a prominent part in the campaign to get GFAVo accepted. The weak counter offensive made by the utilities in late 1982 and early 1983 came too late. The political situation had by then altered radically in favour of anti-coal and pro-forest policies.

The utilities lost the argument because of their growing isolation. Always a target for public suspicion, they received increasingly less support for their views from the coal industry, the trade unions and even the Confederation of German Industry (Bund der Deutschen Industrie-BDI). In 1978, the BDI had complained that government was 'forcing investment in environmental protection as part of its counter-cyclical policy' and called for a halt to Federal enthusiasms for air-pollution abatement. By 1982, however, it accepted the need for stricter environmental regulation, although still arguing that the problem of forest damage was far too complex to be attributed to air pollution alone.

Accused of damaging the environment, the energy industries' initial response was identical to that of their British counterparts. In their opposition to more stringent emission abatement, the electricity industry, the coal-mining industry and BMWi initially relied largely on economic arguments, especially the threatened loss of international competitiveness due to rises in electricity prices.

These arguments were similar to those which won out in the UK. In contrast to what happened in Whitehall, however, the case for more stringent environ-mental regulation had increasingly more powerful supporters in Bonn.

German utilities were not state-owned and, in the early 1980s, were cash-rich, enjoying healthy profits and facing no major capital expenditure programmes. Some argued that their wealth would be well spent on an emissions control programme which would not add to generating capacity, but which would nevertheless stimulate economic activity and technological change. If some older coal-fired power stations were closed down, then this too would encourage the desired rationalization process.

Experts had already warned that, given the spiralling cost of nuclear safety, the cost of nuclear power would rise above that of coal-produced electricity by the mid-1980s.[14] Higher generating costs for fossil fuels would enhance the economics of nuclear power, and could be justified with reference to saving the forests.

In order to prevent a Federal solution, the major electricity utility with deep involvements in both the nuclear and the coal business, RWE, had first tried, in July 1981, to negotiate a 'voluntary' agreement with the government of its home state, NRW. This would have involved an emission standard for SO_2 of 650 mg/m^3 and would have required it to fit a relatively cheap abatement technology, the dry scrubber. The relatively low SO_2 removal obtained from dry scrubbers would not have satisfied the standards required under the 1983 Regulation.[15]

RWE later admitted that the proposed voluntary agreement with NRW would have been tougher than the draft GFAVo agreed by the Schmidt Government in September 1982. RWE had then hoped that existing and lignite-fired plants would be exempt from the toughest provisions of the GFAVo. This persuaded environmentalists to attack the 1982 proposal as 'RWE's Law', tailor-made to the wishes of this powerful and not entirely popular institution.

The BDI, of which the utility trade organisation VDEW, because of its close ties with local government, is not a member, took a different approach. It had realized that a national commitment to clean air would not necessarily be to the disadvantage of its members, although impacts on international competitiveness remained a worry. This concern in turn fuelled the pressures in support of more stringent European environmental standards which the FR Germany was soon to apply to European Community institutions.

In early 1982, the German trade unions abandoned their opposition to more stringent emission control, having condemned earlier drafts of the GFAVo as 'anti-coal'. This adjustment by the trade unions to the needs of clean air was not complete, however, until 1984 when the trade unions as a whole came to perceive environmental protection as stimulating investment and helping to create employment.

Faced with the allegation of environmental blame, which science could neither confirm nor deny, resistance by most German fossil fuel producers had crumbled very quickly. Warned by environmentalists, the coal industry and the mining unions became convinced that coal use would have a future only if it were to develop a clean image through technological fixes. The coal industry therefore 'greened' even before the electric utilities (and some senior politicians) had finally made up their minds.

The new government

The period between October 1982 and March 1983 was decisive in the creation of the political energy needed not only to adopt, but to implement, the German clean-air initiative. In October 1982, Helmut Kohl became Chancellor and announced a U-turn in German politics. However, for air-pollution control at least, there was policy continuity and implementation became more energetic.

When the government changed, there were important new Ministerial appointments to be made. The CDU wished to keep Franz-Josef Strauss, the leader of the Bavarian CSU, at home in Munich. To do so, it was necessary to reach a compromise whereby the FDP gave up some of its posts in favour of other CSU politicians. Both the BMI and the BMELF passed into the hands of Strauss's personal friends. Dr Friedrich Zimmermann, a right-wing Bavarian lawyer fond of hunting and representing many small forest farmers, became Interior Minister. His conservatism severely tested his popularity, something which he tried to counteract by dedicated support for clean-air policy and the campaign to save the German forests. The only ministry critical to the determination of clean-air policy to remain in FDP hands, was the BMWi, the defender of the energy industries.

Zimmermann inherited several serious problems: how to deal with the increasingly vociferous German peace movements; how to respond to the growing likelihood of the Green Party entering the Bundestag; and how to manage nuclear issues, especially waste disposal, more acceptably. Most importantly, he had to create a greener image for the new government in anticipation of the election battle which the new Coalition had to face almost immediately.

The new government, as yet unelected, immediately began a crusade for clean air, the centrepieces of which were the GFAVo, the adoption of US type vehicle emission limits and large increases in R&D funding for forestry and pollution research. The SPD in opposition countered by advocating a general levy on fuel use to raise funds which would fund measures to help the forests, but this was not accepted by Bonn.

The new government also disbanded the Parliamentary Commission of Inquiry on Energy and Environment (the Enquete Commission), which was hopelessly divided over future energy policy because of disagreements over the role of nuclear power.

The 'Save Our Forest Federal Action Programme'

The Federal election took place in March 1983 after a campaign in which the threat of a red – green alliance became one of the key issues. The election confirmed the caretaker CDU/CSU/FDP government by a narrow margin and brought the Green Party into the Bundestag. To the annoyance of the CSU, the Union parties still had to rely on the FDP as a coalition partner. Without the desertion of many voters from the SPD to the Greens, the outcome might have been different.

A programme to save the forests became a platform in the election battle, with the CSU in particular becoming the champion of the forests and therefore clean air. The CSU had for some time expressed concern about forest damage based on expert advice from the Bavarian Environment Ministry. As the CSU had long been the party most dedicated to nuclear power, it had little difficulty in connecting the two issues.

In the election campaign, clean air proved to be a critical issue. No party could safely declare itself to be opposed to a major clean-up of emissions, while the CDU/CSU could undermine protests against nuclear power with reference to its environmental benefits. Energy programmes proposed by the Greens, which foresaw an important transitional role for coal-based electricity, were made to look less credible. Unlike in the UK, the dangers of acid rain were not discounted or even subjected to rigorous scientific testing, but were amplified with official sanction. Soon, almost everybody, and especially the general public, became convinced that German forests needed to be saved from the effects of long-range acid deposition.

In early 1983 the Advisory Council on Environmental Questions (RSU) proposed measures to reduce significantly NO_x emissions from combustion plants. It also urged government not to commit industry to any specific abatement technology for existing plants, but to rely on emissions limits which would encourage industry to develop its own solutions. These recommendations were incorporated in the GFAVo by the new government.

The new air-pollution control legislation was accepted by the Bundesrat in early summer and became law in July 1983, but only after a further debate which led to much tighter standards than those envisaged in 1982 being adopted. A proposal by the government to restrict the applicability of the GFAVo by applying the 50 MW cut-off threshold to individual units at combustion sites was rejected by the Bundesrat. Green experts were alert to proposals which would weaken the GFAVo and quickly gave them wide publicity.

The arguments advanced against the GFAVo in its final form are given some prominence here because they show that the views which came to dominate the debate in Britain, were also expressed in Germany.

Final resistance to the GFAVo

A BMWi working group on energy policy reporting in early 1983 again pointed to the danger of rising energy prices leading to a loss of competitiveness of German exports, the duplication of regulations, reduced competition between fuels and disregard of the principles of market economics. When the VDEW was asked in early 1983 to comment on proposals made by a working group on energy and environment set up by Bonn and the Länder, it approved only six of twenty-five proposed measures, namely those dealing with public financial assistance for environmental protection measures. The idea that BImSchG, or even worse, the Energy Act (EnWiG), should be significantly revised (as had been suggested by the working group) was totally rejected at that time.

In July 1983 a closely reasoned article published in a major energy journal hinted at the possibility of a constitutional challenge to the GFAVo, questioning both the involvement of SO_2 in forest damage and the efficiency of the entire clean air programme.[16] This programme was seen as part of an 'anti-coal' strategy promoted by the pro-nuclear government, with an increase to 80 per cent in the nuclear contribution to total electricity generation allegedly being the official intention. It was argued that the whole exercise would remain ecologically pointless unless the whole of Europe were forced to adopt a similar strategy. Echoing VDEW, the article suggested that the TA (Luft) should become a proper, legally binding instrument and that there should be more official emphasis on cost-benefit thinking.

Other measures

In December 1983, official BMI publications called for emergency measures against forest dieback. The 21st Conference of the Länder environment ministers recommended the introduction of lead-free petrol from January 1986 in order to pave the way for the introduction of catalytic convertors and US-style vehicle emission standards. It argued that 'drastic limits to the burden placed on the environment combined with securing of an assured supply of energy, should be the all-embracing aim of policy'.[17] The major energy-environment initiative was under way with the full support of the Länder.

The Greens and the clean air programme

Green supporters and writers helped to ensure that the acid-rain issue was widely debated and discussed in Germany, as is shown by the popularity of low priced paperback books dealing with the subject. The environmental education of the German public was continuing. Many of these books challenged Government policy from a Green perspective, linking the forest problem to a transformation of the energy economy, i.e. to energy efficiency and reduced consumption in general by a profligate industrial society.[18]

In 1983, it was politically impossible for the Greens to oppose measures which were designed to save forests. They had assisted the BMI by insisting on a debate on emergency measures for cars, including speed limits. The idea of a 100 kilometre/hour speed limit turned out to be quite unacceptable to the car industry, however, because it saw high performance not only as a sign of quality but also as a major stimulus to engineering advances. Drivers were also resistant to the prospect of losing the right to travel, on certain stretches of road, as fast as their cars would go. This argument continues today as the energy and environment debate itself assumes new dimensions.

Once in the Bunderstag, the Green Party began flooding legislators with proposals for change: by July 1984 they had suggested 22 draft bills (out of a total of 60 from all parties) and had inundated government ministries with formal parliamentary questions. The Greens strongly promoted renewable energy sources and energy efficiency, ideas for which they could at the time find little support. However, by the end of the decade, some green ideas gained wider currency as economists argued that energy prices should rise in order to reflect the external environmental costs of energy consumption.

The death of forest dieback

The great national crisis (though not the ecological problem) of Waldsterben effectively ended in 1985 when the Federal government informed Parliament that 'the rapid increase in forest damage observed since 1982 has not altogether continued . . . two thirds of the damaged areas fall into category one', i.e. were weakly damaged.

The term Waldsterben, forest dieback, disappeared from the official vocabulary and was replaced by Waldschäden, forest damages. Four distinct types of forest damage have now been identified: direct damage by high concentrations of SO_2; needle yellowing and loss at medium to high altitudes; needle reddening of older trees followed by needle loss and fungi attack; and needle yellowing on calcareous soil in high altitudes. Damage affecting the trees on Alpine slopes remains most poorly understood.

As described in Chapter 3, the precise role played by air pollution remains uncertain, although there is now some scientific consensus that forest damage is a general stress symptom which results from a growing imbalance between nutrients available in the soil (which are reduced both by acid leaching and by repeated harvesting of wood), and growth-stimulating nutrients contributed by air pollution. The causality is complex and is not yet fully understood.

In 1986, the newly formed Federal Environment Ministry (BMU) put together rather hurriedly by Kohl, reported the conclusions of its expert advisers as follows:

> there is no single type of forest damage and no single cause. We are dealing with a highly complex phenomenon which is difficult to untangle and in which air pollutants play a decisive role.

Only the future will tell whether the great clean-up of the air in the 1980s will have the hoped-for effects on forest ecology, human health, freshwater life and buildings. It will certainly not have harmed any of these targets.

The Green Party and environmentalists are not yet satisfied. They believe that an SO_2 emission limit of 100 mg/m^3 could be achieved by state-of-the-art technology rather than the 400 mg/m^3 limit contained in the GFAVo. Together with the SPD, the Green Party wants to see the BImSchG amended to impose a legal duty on emitters to use the 'best state-of-the-art technology'.

However, air pollution and, to a growing extent, the global warming problem, continue to be politically useful to those who wish to attack the Green's environmental policies. The BMU Minister has reminded opponents of nuclear power that:

> those of you who are in favour of shutting down nuclear power stations, cannot complain one day about forest damage and major air polluters and the next demand the immediate replacement of nuclear power stations by coal burning ones.[19]

Implementation: 1983–89

Legal perspectives

The enactment of the GFAVo still left some legal ends to be tied up. In particular, a legal challenge to the constitutionality of the retrospective requirements to clean up existing combustion plants was still possible. Further legal underpinning was required.

The BImSchG was amended in 1985 to remove the possibility of a constitutional challenge to the GFAVo, as described in Chapter 9. In addition, the TA (Luft) was amended in 1986 to apply stringent emission standards to virtually the entire industrial and commercial sectors. This step prevented the possibility of higher sulphur fuels being driven out of the large combustion sector by the GFAVo and dumped in other markets. A small combustion plant regulation covers household emissions.

The implementation of the GFAVo

Operators of large combustion plants were required to decide, by July 1984, how they proposed to comply with the GFAVo requirements. For existing plants, the choice was between retrofitting plant or closing it down.

The utilities decided to comply by fitting FGD, a technology already commercially available in Japan and the USA, to most power stations. Although technology licensed from abroad was used, adaptations were necessary to take account of specific operating conditions and fuels.

For NO_x, however, the utilities had to carry out more fundamental development and demonstration work to apply technologies used so far only in Japan to German operating conditions. The established technical means for reducing emissions, combustion modification, was not considered to be state-of-the-art. The Bundesrat had tightened the NO_x emission limits in the GFAVo to make selective catalytic reduction (SCR) of NO_x essential for compliance. In 1988 over 60 pilot denitrification plants were in operation and 12 000 MW of capacity was in place. By 1993, 30 000 MW of hard-coal fired plant will have been retrofitted with SCR. Brown coal-fired plants have been required to use only combustion modification.

The retrofitting of coal-fired power stations with FGD is considered to have been a complete success. By 1988, FGD had been fitted to 26 000 MW of hard coal capacity and 10 000 MW of brown coal-fired plant. Some 12 500 MW of small hard coal and brown coal-fired power stations have been shut down. The remarkable speed with which the FGD retrofit programme has been carried out reflects a number of factors, including: the perceived urgency of the forest damage problem; the availability, in the early 1980s, of spare industrial capacity which could undertake the massive construction programme;[20] and the fact that many German power stations run at relatively low annual utilization rates with annual downtimes for maintenance, providing ample opportunities to retrofit.

Practical concerns arising from FGD include the associated waste disposal problems; 1.6 million tonnes of such wastes are now produced annually in NRW alone, compared with 85 000 tonnes in Bavaria. Some wastes are converted to gypsum or sulphuric acid, while others are dumped in disused mines.

The total cost of cleaning up power stations is estimated to be DM 21bn, of which DM 14.3bn is for retrofitting FGD and DM 7bn is for retrofitting SCR. In comparison, the estimated damages arising from air pollution annually remain much more controversial, the estimated range being DM 10–40bn. It is now estimated by the utilities that nuclear power is a cheaper form of power generation than brown coal stations because of the imposition of controls on fossil fuel emissions. Bayernwerk AG and PreussenElektra, the two utilities most reliant on nuclear power, offer the lowest electricity prices to large customers.[21]

By 1990, the electricity supply industry had over-achieved by 10 per cent its 1993 goal of a 75 per cent cut in SO_2 emissions. NO_x emissions had been reduced by 50 per cent of their 1982 value.[22] By the early 1990s, NO_x reductions from power generation are expected to reach 70 per cent.[23] Germany is therefore very much ahead of the compliance requirements for the 30 per cent Protocol to the LRTAP Convention and the EC's LCP Directive, both in terms of the size of emissions reductions and the dates of the achievement. The objective of a 30 per cent reduction overall for NO_x has not yet been achieved, however, because of increases in transport emissions.

Car emissions and catalytic converters

In 1983, the German car industry was required to prepare for a full conversion to lead-free petrol and catalytic converters. The industry, however, successfully argued for significant delays which it needed to adjust without serious commercial implications.

Total emissions of NO_x have nevertheless kept on rising. The voluntary emission control measures which the government was able to adopt under EC rules have remained insufficient, because of overall growth in road traffic,[24] a low replacement rate of old cars and because heavy vehicles still remain largely uncontrolled.

The car industry is in part responsible for this delay. It decided that the potential commercial impact of unilateral action was too risky and thus adopted successful delaying tactics by successfully demanding a European solution (see Chapter 13). The risk of German motorists being unable to obtain suitable petrol while driving abroad proved to be a convincing argument. The vehicle industry also encouraged many motorists to buy diesel-powered cars, because these complied with the gaseous emission standards agreed in Brussels in 1985.

The car industry therefore pleaded successfully that the needed technology change required time for the establishment of both a proper fuel supply system and the adoption of new vehicle designs appropriate for catalytic converters. A demand for more expensive, if cleaner, cars had also to be created. For this industry was able to rely, in Germany, on substantial government help.

Research and development

The BMFT has been promoting research into pollution abatement technology since the first half of the 1970s. It currently supports 590 research projects on the causes of forest damages costing DM 277m. In August 1988, the government announced the continuation of its Action Programme to Save the Forests, noting that 50 per cent of forests still suffered from reduced vitality and 15 per cent from moderate and serious damage. R&D of both a scientific and technological nature benefits from this programme.

Economic incentives

In 1989, the Hamburg Institute for Economic Research estimated that 3 per cent of GNP was spent on servicing capital and operating installations for environmental protection, a rise from 1.9 per cent in 1975.[25] However, the objective of keeping public expenditure as low as possible has not been one of the foundations of environmental policy in Germany. In September 1987, the new Environment Minister, Professor Töpfer announced that a total of DM 50bn of Federal funds were to be 'set in motion' in order to assist in the implementation of a Federal air-pollution control package.

Economic incentives to persuade consumers to buy environment-friendly products, and especially new cars, have been adopted to the satisfaction of the car industry. Tax rebates and tax relief (up to DM 2200 for new large cars – with engines greater than two litres – and DM 1000 for small cars) have persuaded car owners to buy new clean cars requiring lead-free petrol or to retrofit their old cars.

Leaded two-star petrol became unavailable in early 1986 making the switch to Euro-norm petrol fairly painless. By mid-1989, over half of the petrol sold was unleaded and being sold in three varieties. About a third of the car fleet complies with the 1985 EC standards, but not yet the higher US-type standards which apply only to large cars. In 1989, large cars made up about 30 per cent of the German fleet.

These achievements were considered inadequate and, in early 1990, extra incentives for clean, small cars and a new system of vehicle taxation based on emission characteristics rather than engine size were adopted.

The clean-up of power stations was achieved with little support from the public purse, although there have been major exceptions. The Buschhaus power station is the best known example, with Bonn contributing DM 260m to the cost of building the regenerative FGD system installed there.

To ensure the acceptability of the entire clean-air package, covering not only power stations but also the entire industrial plant stock, government also paid special attention to those firms which were either unable to pay for abatement or which would be seriously disadvantaged. In such cases the authorities could appeal to the 'Community Burden' principle (see Chapter 9). Bonn and the Länder implemented numerous provisions involving tax relief and other economic incentives, including low-interest loans under the Economic Recovery Programme (ERP – a left-over from the Marshall aid plan), and a whole host of other existing and new programmes. These were more relevant for medium- and small-sized firms and played a major role for environmental protection measures other than acid-rain abatement.

Even these direct subsidies are perceived less as intervention than as promotion or lubrication of the market so that it can be guided towards more pollution abatement through technological innovation and investment. For environmental protection in general, the amount of tax relief officially granted to small and medium-sized business rose from DM 1.1 billion in 1981 to 3.8 billion in 1985. As mentioned in Chapter 7, this method of economic management is widely used in the FR Germany.

The European dimension

All parties in West Germany agree that the economic burden of environmental protection should be widely shared. This has had implications for Britain and the other EC Member States. The BDI and other bodies argue that the German Federal Government should attempt to 'spread the misery' of stringent environmental controls to other European partners in order to protect German industry against a potential loss of competitiveness. A legal basis for doing this exists in the EC through its commitment to the harmonization of legislation impinging on trade and, more recently, the co-ordination of national environmental policies.

Uniform environmental standards will undoubtedly remain a part of the package which Germany expects from the Single Market. German activity at the EC level is interlinked with its domestic environmental policy to the extent that the rest of Europe is expected to follow the German example. So far, Bonn has been successful with regard to vehicles. The attempt to harmonize power station controls, because of the very different conditions of Member States, has been less successful. The European negotiations on the Large Combustion Plant Directive are described in Chapter 12, following the British story.

Notes

1. S.A. Boehmer-Christiansen (1989), *The Politics of Environment and Acid Rain: Forests Versus Fossil Fuels*, SPRU Occasional Paper 29, Brighton.
2. E. Müller (1984), *Die Innenwelt der Umweltpolitik*, Westdeutscher Verlag, Opladen; and H. Weidner (1986), *Air Pollution Strategies and Policies in FR Germany*, Wissenschaftzentrum, Berlin.
3. P. Davids (1984), *Die Grossfeuerungsanlagenverordnung: Technical Kommentar*, VDI Verlag, Düsseldorf.
4. S.A. Boehmer-Christiansen (1990), The regulation of vehicle emissions in Europe, *Energy and Environment*, Vol. 1, No. 1, 1–25.
5. 'Politik' means both policy and politics, e.g. Umweltpolitik, refers to both policy and its making through political processes.
6. E. Müller (1984), *Die Innenwelt der Umweltpolitik*, Westdeutscher Verlag, Opladen.
7. Ibid., p. 136.
8. H. Weidner, 17 November 1989, Die Umweltpolitik der konservativ-liberalen Regierung, *Aus Politik und Zeitgeschichte* (Beilage zu Das Parlament), Bonn.

9. 16 November 1981, Das stille Sterben Der Saure Regen zerstört den deutschen Wald, *Der Spiegel*, **35** (47), 96–110.
10. BMFT (March 1985), *Umweltforschung zu Waldschäden*, Third Report, Bonn.
11. L. Wicke (1986), *Die Ökologischen Milliarden*, Munich.
12. E. Müller (1984), *Die Innenwelt der Umweltpolitik*, Westdeutscher Verlag, Opladen.
13. H. Weidner (17 November 1989), Die Umweltpolitik der konservativ-liberalen Regierung, *Aus Politik und Zeitgeschichte* (Beilage zu Das Parlament), Bonn.
14. U. Hansen (1984), Nuclear economics in the Federal Republic of Germany, Paper presented to the *Sixth International Conference of the Association of Energy Economists*, Cambridge, 9–11 April 1984.
15. L. Mez (September 1984), *Neue Wege in der Luftreinhaltepolitik: Eine Fallstudie der RWE*, IIUG Report 84–3, Wissenschaftszentrum Berlin.
16. E. Gerking (1983), Die GFAVo – ein Fall für Karlsruhe, *Energie-wirtschaftliche Tagesfragen*, 35.
17. BMU (1983), *Umwelt 99*, Bonn, p. 4.
18. Öko-Institut (April 1984), *Rettung für den Wald: Strategien und Aktionen*, Fischer alternativ.
19. BMI (November 1986), *Umwelt*, Bonn, p. 5.
20. In particular, because of a downturn in the shipbuilding industry, there was a high availability of labour skilled in welding which was one of the most important activities involved in fitting the long flue gas ducts necessary at existing power stations. This factor would have played some role in BMWi's assessment of the desirability of the clean-air programme. Personal communication, B. Schärer, Umweltbundesamt.
21. VDEW (March 1990), *Stromthemen*, Frankfurt/Main, pp. 1–2.
22. Information Office of the Electricity Industry (March 1990), *Stromthemen*, 7, 3.
23. Financial Times Business Information (26 May 1988), *Power in Europe*, No. 25.
24. H. Westheide (1987), *Die Einführung bleifreien Benzins in der BR Deutschland mit Hilfe okonomischer Anreize*, Erich Schmidt, Berlin.
25. *Wissenschaft, Wirtschaft und Politik* (January 1989).

11 Science and money: the British response

Introduction

This chapter describes the evolution of British policy on acid rain throughout the 1980s. The story told is very different from the German one. German strategies developed within a pluralistic decision-making system, with a strong Federal bureaucracy providing input to a dynamic political process. In Britain, party-political involvement in environmental issues has, until recently, been virtually absent and the relevant bureaucratic institutions, both those within government itself and those charged with implementing policy, have been weak, under-resourced and often unassertive. As a result, most of the major acid-rain decisions have been taken at a high level without the influence of complex, countervailing pressures deriving from well-established bureaucratic and regulatory institutions which operated in Germany.

At the same time, public concern about environmental issues was diffuse and rather shallow throughout most of the 1980s, while there existed a much more sharply focused set of industrial and institutional concerns, notably in the energy sector. Acid-rain abatement did not appear to offer political rewards since controls would have frustrated rather than aided political objectives which enjoyed a higher priority. Nevertheless, international demands for acid emission controls grew more pressing as time passed.

British policy-making can be divided into five chronological stages. The first covers the period 1972–83, from the Stockholm Conference on the Human Environment through to the point at which international pressures began to take on a more urgent nature.

During the second period, 1984–5, it became necessary for the government to establish an unambiguous policy line on acid rain. It was decided that no expensive mitigative action was justified because of the inconclusiveness of scientific evidence.

During 1986–7, the third period, a limited programme of acid emission abatement was formulated, on the initiative of the Central Electricity Generating Board (CEGB) rather than the government. This move was justified with reference to the results of scientific experiments in Scandinavia.

In 1988, the UK was forced into an undertaking to establish more ambitious plans to abate sulphur dioxide (SO_2) and nitrogen oxides (NO_x) to a much greater extent than the electricity supply industry wished. This undertaking was established just as plans for electricity privatization were getting under way.

During the fifth and final period, covering 1989–90, the Department of the Environment (DoE) began to formulate plans for implementing European obligations. Whereas, in Germany, the central interests of the coal and electricity industries had been a major consideration in the formulation of acid emission abatement policies, these issues were given a full airing only at the implementation stage in the UK when fundamental differences between the interests of the coal and electricity supply industries emerged.

This chapter is concerned primarily with the development of policy on emissions from stationary sources. Throughout the 1980s, the UK was also an important participant in the European debate about vehicle emissions regulation. There are many similarities between the positions which the UK adopted for vehicles and power stations. In particular, the UK attempted, for much of the 1980s, to defend lean-burn engine technology, which had been favoured by British Leyland, against tight vehicle emission standards which would have required the use of more expensive three-way catalytic convertors. As for power stations, the UK was required to modify its position considerably by the end of the 1980s.

Britain's early position, 1972–83

The attention of most European governments, including that in the UK, was drawn to the acid-rain problem in 1972 prior to the Stockholm Conference on the Human Environment. The evidence presented by Scandinavian scientists linking acid rain and emissions of SO_2 was given recognition in the official Conference Declaration. The UK government did, at the time, acknowledge the plausibility of this link. A report commissioned by the DoE in preparation for the Stockholm Conference observed that:

> International agreement [on SO_2] is therefore desirable, both to control the distribution of low sulphur fuels, and to ensure that the emission of SO_2 from high chimneys in one country does not cause increased pollution elsewhere.[1]

However, fifteen years passed before the UK was to enter into any international commitments. The 1973 oil crisis did much to diminish the perceived urgency of air pollution which began to be taken seriously again only in the late 1970s.

From 1972 onwards, environmental groups, politicians and scientists in Scandinavia exerted a persistent, if ineffectual pressure on Britain to take measures to reduce SO_2 emissions. Much of that pressure was targeted on the CEGB, as its 'tall stacks' policy was seen as exacerbating the problem by injecting SO_2 high into the stratosphere where it could be carried hundreds of miles before being deposited in the form of acid rain.

The CEGB, as the state-owned utility responsible for the bulk of Britain's SO_2 emissions and with a long-standing statutory responsibility to have regard for the environmental impacts of its activities, was an appropriate target for direct bilateral pressures. However, as described in Chapter 2, Scandinavian governments also took the multilateral approach, using the machinery of the UN Economic Commission for Europe (UNECE).

The CEGB's strategy

The early response from the CEGB was twofold. First, it voiced scepticism about the conclusiveness of scientific evidence on acid rain and emphasized the lack of proven, direct causal links between SO_2 emissions and environmental damage. The arguments tossed back and forth between the CEGB and its critics were of the nature of those described in Chapter 3.

Second, the CEGB argued that the cost of installing flue gas desulphurization (FGD) technology on its power stations was excessive. All of the abatement equipment would need to be installed at existing power stations, where the fitting of these major chemical plants would be difficult and expensive. The CEGB did not believe that the evidence available from Scandinavia was conclusive enough to warrant the expenditure of large sums of money, or that, even if it wanted to, it could obtain Treasury sanction.

After the oil crisis, the CEGB also subscribed to the general view which emerged within the UK energy establishment that the SO_2 problem would vanish during the 1990s and beyond as traditional coal-fired power stations were replaced by nuclear reactors and coal-fired stations using pressurized fluidized bed combustion (PFBC) technology which would give rise to intrinsically lower SO_2 emissions. Thus, FGD was seen merely as advancing emission reductions which were likely to take place in any event in the longer term. The reasons for CEGB mistrust of FGD as a technology were described in Chapter 8.

If it had been possible to persist with this strategy, the acid-rain problem would effectively have gone away as new, less polluting power stations were installed. The CEGB's basic policy remained in place under three different chairmen between 1973 and the mid-1980s.

In the early 1980s, the National Coal Board (NCB) adopted very much the same position as the CEGB, its principal customer. However, in light of subsequent developments in 1989–90, it might well have been in the longer term interest of the NCB to have argued for the installation of FGD in order to establish markets for high-sulphur British coal which would have been resilient to environmental controls.

UK science and acid rain

The most tangible outcome of the CEGB's policy was a programme of acid-rain research, designed to build up internal scientific expertise. The results of this research were generally deployed in a defensive mode to deflect criticisms of the Board's environmental performance. Impressive and expensive programmes were built up in biology, in the earth sciences and in atmospheric physics and chemistry, with the aim of understanding the transport and transformation of power-station flue gases and their environmental impacts.

Table 11.1   Research on acid deposition in the UK 1972–84
(£k per year)

	WSL[1]	DoE, of which AERE[2]	Others	NERC[3]	Met Office	CEGB
1983–84	200	220	209	n.a.	1300	1146
1982–83	200	120	28	860	1360	1401
1981–82	174	115	163	691	1383	1035
1980–81	147	115	163	550	918	1155
1979–80	92	115	152	372	1026	1462
1978–79	56	90	69	273	825	596
1977–78	21	95	65	350	n.a.	339
1976–77	20	110	–	191	n.a.	215
1975–76	41	132	–	158	n.a.	83
1974–75	43	118	–	n.a.	n.a.	50
1973–74	44	93	–	n.a.	n.a.	n.a.
1972–73	10	105	–	n.a.	n.a.	n.a.

Notes:
(1) Warren Spring Laboratory;
(2) Atomic Energy Research Establishment;
(3) Natural Environment Research Council;
(4) n.a. = not available.

Source: Department of Environment.

The two former programmes have focused on the changing chemistry of soils and the factors affecting the mortality of fish in acidified waters. Later, in the 1980s, work to assess the phenomenon of forest dieback was added.

In 1978–9, the CEGB spent some £1.5m per year on acid rain research, 45 per cent of the total expenditure in the UK. The Meteorological Office accounted for most of the remainder. Acid-rain research expenditure in the UK over the period 1972–84 is shown in Table 11.1.

Such programmes could have been built up only in an organization such as the CEGB, with its public-service ethos. A private-sector utility in a country as small as Britain could not have mounted these programmes. Nor could they have been built up so quickly and with such a sense of purpose in other parts of the public sector in the absence of any urgent domestic environmental threat. The Natural Environment Research Council took a decade to build up its acid-rain research budget to levels approaching those of the CEGB.

The CEGB was sensitive to criticisms that its research might be biased. Its most ambitious and visible research effort was not therefore carried in-house. In 1983, along with the National Coal Board (NCB), it sponsored the £5m Surface Waters Acidification Programme (SWAP) which was run by the Royal Society in concert with the Swedish and Norwegian Academies of Science. This programme was intended to produce results which would give indications of appropriate policy responses to the acid-rain problem over the period 1986–8. However, even this arrangement ran into public-relations difficulties as some

Scandinavian scientists viewed the research funds as a 'bribe in order to get out of our [British] commitments'.[2]

The CEGB ran into further criticism in 1985 when a video film which it had commissioned on the subject of acid rain was said by Norwegian scientists to have 'concealed and twisted' facts.[3]

The DoE itself later increased its research efforts on acid rain. In particular, after 1980, it established four independent Review Groups (covering Acid Deposition, Acid Waters, Terrestrial Effects and Building Effects) to assess available data and advise on research needs in relation to acid rain. The first report of the Acid Deposition Review Group appeared in 1983 and the other groups have reported since.[4] The Second Report of the Acid Waters Review Group in December 1988 was particularly unambiguous about the substantial reduction in SO_2 emissions which would be required if environmental targets, such as the prevention of further deterioration of acid waters, were to be met.

Technology research

As well as pursuing scientific research themes, the CEGB and the NCB also funded research and development (R&D) work on technologies which, if commercialized, would reduce acid emissions. In particular, considerable amounts of money were spent on the development of PFBC technology at the large-scale experimental facility at Grimethorpe. Within the larger picture of R&D spending, work on nuclear technology continued to dominate. However, little effort was expended on technologies, such as FGD, which offered more immediate prospects of SO_2 abatement. Indeed, the CEGB's view was that such technologies were, in any case, fully commercial by 1980.[5]

The first detailed cost study of fitting modern FGD to UK power stations was carried out jointly by the CEGB and the NCB only in 1982–4, and then at the suggestion of the Advisory Committee on Research and Development (ACORD).[6] The CEGB also carried out relatively low-cost work on a pilot installation of a hydrogen chloride pre-scrubber (necessary because of the high chlorine content of British coal), low-NO_x burners and, with the NCB, the possibility of enhanced washing of sulphur from coal.

Thus, the technology research portfolio of the UK energy industries was structured towards a long-term response to the problem of acid emissions as existing generation technologies became obsolescent. The CEGB was not well placed to respond rapidly should reductions in SO_2 emissions have been deemed politically desirable.

The international dimension

International interest in transboundary air pollution re-emerged in the late 1970s after the first effects of the 1973 oil crisis had faded and negotiations on global marine pollution reached their conclusion. The UK signed the Geneva Convention on Long Range Transboundary Air Pollution (LRTAP) agreed in 1979, but fought against the inclusion of any specific commitments to stablize or reduce SO_2 emissions. As will be described below, the UK also declined to sign a subsequent Protocol to the Convention which required countries to reduce their SO_2 emissions by 30 per cent.

Scandinavian and Canadian pressure to give the LRTAP Convention teeth through more ambitious protocols continued in the period 1979–83. The 1982 Stockholm Conference on the 'Acidification of the Environment' renewed pressures on countries, such as the UK and the US, which were reluctant to abate SO_2 emissions. It was at this conference that the FR Germany performed its U-turn and announced that it was joining the 30% Club.

By early 1984, the CEGB and the government were still putting up a vigorous defence of their policy of inaction within a cloak of scientific reason:

> If it is shown that emissions from our power stations cause significant damage in the environment, whether in the UK or elsewhere, and if reducing these emissions would both solve the problem and represent the most cost-effective way of doing so, then the CEGB would take whatever action was necessary, even though this would increases the cost of producing electricity.[7]

The dirty man of Europe 1984–85

Renewed pressures

The CEGB's and the government's approaches were put to the test in 1984 when acid rain became a headline issue in Britain. In December 1983, the European Commission put forward its Large Combustion Plant (LCP) Directive,[8] proposing substantial reductions in emissions of SO_2, NO_x and dust from power stations and larger industrial boilers and furnaces. Meanwhile, in March 1984, the Scandinavian countries succeeded in having a draft Protocol to the LRTAP Convention drawn up, which would require signatories to reduce their SO_2 levels by 30 per cent between 1980 and 1993.

Britain thus found itself facing two major international initiatives which profoundly challenged its established policy position. At the same time, environmental groups stepped up their campaigns for a reduction in SO_2 and NO_x emissions, characterizing Britain as the 'dirty man of Europe'. Led by the Scandinavian non-governmental organizations (NGOs), and with the participation of Friends of the Earth in the UK, an international Stop Acid Rain Campaign was active in 1984–5. This was to have no perceptible effect on UK policy.

A wide range of individuals and organizations published reports on acid rain during this period. The consultant's report which had underpinned the EC Commission's draft LCP Directive was published in a popular format and was much quoted by politicians and the media.[9] The Nature Conservancy Council, a statutory body, and Earth Resources Research (a 'green' consultancy frequently used by Friends of the Earth and Greenpeace) both published reports raising the possibility of acid-rain damage in the UK.[10] Friends of the Earth produced a particularly influential report on forest damage[11] which forced the Forestry Commission to examine the possibility that air pollution was damaging British trees.

A more cautious report was published by the Watt Committee on Energy, a body representing professional bodies and interests.[12] A number of paperbacks aimed at a wider audience also appeared.[13]

Industrial responses

It was not only the CEGB and the NCB which were opposed to acid-emission controls. The Confederation of British Industry (CBI), representing a wide range of industrial interests, also argued that it would be 'premature and unwise to make costly legislative demands'.[14] However, the CBI were generally supportive of the idea of regulating new plant, while expressing strong scepticism about the need to regulate plant as small as 50 megawatts (MW) heat input as proposed by the European Commission. The CBI's arguments in favour of a 100 MW cut-off point for emissions regulation were to become an important, though perhaps disproportionate, element of negotiations on the LCP Directive. The CBI's main channels of influence on government were through the Department of Trade and Industry (DTI) and the DoE.

Parliament and acid rain

Acid rain also attracted the attention of politicians. The House of Lords Committee on the European Communities reviews all European legislation and held an Inquiry into the industrial air pollution 'framework' directive and the proposed LCP directive in Spring 1984.

The Committee endorsed the general view held in the UK energy establishment that the long-term development of PFBC was the most promising route for reducing SO_2 emissions, but did recommend the retrofitting of FGD to two UK power stations while PFBC was developed.[15] The Committee's sternest criticism was reserved for the European Commission, however, which it accused of 'failing to give a more objective and scientifically sound basis for proposed controls'. The Committee also attack the philosophical basis for the proposed LCP Directive, arguing that 'it would distort not equalise conditions of competition between Member States'. However, on the question of vehicle emissions, the House of Lords Committee warned the government that support for the cheaper lean-burn engine option was unlikely to be tenable.

While the Lords placed themselves somewhere between the UK government and the EC, the House of Commons Environment Committee generated a much more forthright set of conclusions a month later in a major report focused specifically on acid rain.[16]

The responsibility of the Environment Committee is to shadow all the activities of the DoE, including housing and local authority affairs. During the 1979–83 Thatcher administration, these were the issues upon which the Committee chose to focus. However, the problems of securing agreement between Labour and Conservative members on these types of issues had meant that the Committee's work had achieved little impact. As the second Thatcher administration started, a conscious decision was taken to re-focus the Committee's activities on natural environment questions, where the scope for cross-party agreement was greater.

The new Committee included a number of members with strong environmental interests. The new Chairman, Sir Hugh Rossi, who had been a junior Northern Ireland minister until the summer of 1983, added political weight to the Committee.

For the acid-rain inquiry, the Committee visited the CEGB's Leatherhead laboratories, scientific research centres in the FR Germany, Norway and Sweden, as well as sites in Scotland and the Lake District. After viewing acid-rain damage for themselves, and taking account of 100 written submissions and 10 sessions of oral evidence involving the CEGB and the government as well as scientists and environmental groups, the Committee's views were forthright.

Their report, published in July 1984, noted that members were 'deeply disturbed over the United Kingdom's current policy position on acid rain'. Concerns were expressed over the UK's increasing international isolation and, with the CEGB specifically in mind, the way that 'manufacturing and energy-producing industry have used the reticence or uncertainty of scientists to push to the full their case for not embarking on large expenditures on emissions control.'

In effect, the Committee recommended the complete reversal of the UK's acid rain policy:

> the UK . . . should, with its EEC partners, agree an overall level of reduction . . . All new plants should meet SO_2 emission levels contained in the Draft Directive . . . the UK join the 30% Club immediately, and . . . this target be achieved by the CEGB being required to reduce its SO_2 emissions accordingly.

The Committee's conclusions were widely reported and its report was perhaps the most influential political counter-balance to government policy coming, as it did, from a cross-party committee.

Wider political views

The political pressures on the government to act on acid rain did not run along either party or ideological lines. The unfashionable 'wet' wing of the Conservative Party, represented by the Bow Group of MPs, which recommended strict power station emission controls in June 1984,[17] might have been expected to attack the government's policy. However, the right-wing Conservative think-tank, the Centre for Policy Studies, co-founded by the Prime Minister in 1974, launched an even more explicit attack, arguing that the lack of action on acid rain was a 'lamentable failure of imagination, as well, possibly of judgement' and that the Conservative Party should 'exproporiate the green issues'.[18]

With an unassailable majority in the House of Commons, the government's supporters did not feel any inhibitions about criticizing its environmental policy.

The government's lack of action on acid rain was vigorously attacked by several speakers at the Conservative Party's 1984 Annual Conference. On the other hand, many Labour MPs representing mining constituencies were suspicious of the extent to which acid-rain controls would undermine the British coal industry.

Given the generally wide support for acid-rain controls among the broader political establishment in 1984, it seems remarkable that a change of government policy did not follow. To explain this, the broader thrust of government policy must be considered.

Environment policy in context

As described in Chapter 5, the DoE cannot make environmental policy in isolation. Regulations which it makes have consequences for private industry, for the nationalized industries, for their departmental sponsors within government and for national income and expenditure. Consequently, the DTI, the Department of Energy (DEn) and the powerful Treasury all had important stakes in the acid-rain controversy. Three specific elements of government policy in the early 1980s constituted obstacles to any substantial moves to curb acid emissions.

DEREGULATION

The first was the general objective of 'rolling back the state'. Implementation of the LCP proposal in particular, with its comprehensive set of emission limits for combustion plant outside the electricity supply industry, would have required a greater allocation of resources for formulating environmental policy and implementing controls. This would inevitably have cut across the government's broader desire to cut bureaucracy and reduce the regulatory burden on industry made evident in the 1986 Efficiency Scrutiny Report on industrial inspection.[19]

MACRO-ECONOMIC POLICY

Macro-economic policy might also have been compromised by major emission controls, particularly if these involved the installation of FGD on large numbers of power stations. The cost of implementing the draft LCP Directive was estimated to have been about £1.4bn in 1984 money.[20] These costs would have fallen almost entirely on the capital expenditure programme of the state-owned CEGB and would, therefore, have counted as part of the Public Sector Borrowing Requirement (PSBR). During the 1980s, the electricity supply industry has, under the financial targets and external financing limits set for it by the Treasury, run down its debt and made a negative contribution to the PSBR. Since reducing the PSBR was a central element of macro-economic policy, acid-rain controls would inevitably have been opposed by the Treasury unless convincing arguments to the contrary had been offered by both the DoE and the DEn.

The privatisation of the electricity supply industry also impinged on acid-rain policy, but at a later date.

ENERGY POLICY

Britain's key decision on acid rain was taken in the middle of the year long miners' strike and while a lengthy Inquiry into the construction of Britain's first Pressurised Water Reactor (PWR) was under way. Joining the 30% Club is reported to have been regarded as 'politically insensitive' during the miners' strike and any reference to the nuclear programme would risk being seen as 'pre-empting the outcome of the public inquiry into the proposed PWR at Sizewell.'[21]

The decision was thus taken at a time when the coal and electricity industries were operating in a highly charged political environment and the CEGB was the prime instrument of government policy in the energy domain. The views of the energy industries were therefore crucial in determining UK policy. The CEGB's own views, informed by an in-built prejudice against the FGD technology and scepticism about scientific evidence concerning acid-rain damage, were well established.

To explain the political sensitivity in 1984, it is necessary to describe the background to the government's coal and nuclear policies. As described in Chapter 8, the electricity supply industry is heavily dependent on British coal as a fuel source. The Conservative government was particularly sensitive to the possibility of industrial action by the National Union of Mineworkers (NUM) leading to interruptions in coal supply and hence power cuts. The previous Conservative government, in which Mrs Thatcher was a junior minister, had effectively been brought down by the six-week national coal strike in 1973–4. This sensitivity was increased following the election of the militant Arthur Scargill to the Presidency of the NUM in 1981. Two policies which would reduce vulnerability to industrial action in the coal industry, one short-term, the other long-term, were put into effect.

The shorter-term policy was the building up of coal stocks in order to resist any NUM strike for as long as possible. A Cabinet Committee was established in 1981 to oversee this process and to devise more extensive contingency plans.[22] In 1981–2, the CEGB was a reluctant partner in government policy. It co-operated in building up power-station coal stocks only after adjustments were made to the CEGB's external financing limits.[23]

While coal stocks were still being built up, the government took great care to avoid provoking any industrial action in the coal industry. In February 1981, the government countermanded an NCB plan to close up to 50 pits after unofficial strike action had swept the coal industry. At the same time, the government persuaded the CEGB to cut back coal imports in exchange for an adjustment to the price of NCB coal.[24]

In March 1984, the closure of the Cortonwood colliery was announced, followed closely by national NUM approval of strikes in Scotland and Yorkshire. NUM leaders avoided a national ballot and hoped that the strike would spread to other regions through a 'domino' effect.[25] However, the strike did not win full support, particularly in Nottinghamshire. The output from working mines, coupled with the high initial level of coal stocks and extra-ordinary measures taken by the CEGB, including increased oil burn and over-firing of power-station boilers to squeeze out more power, meant that no power cuts took place until strikers' morale broke. In March 1985, the NUM called off the strike.

Acid rain was an issue to which the miners' political supporters were acutely sensitive. Some Labour MPs are reported to have said at the time that anyone raising the issue of acid rain was a 'class traitor'[26] and joining the 30% Club could easily have been perceived as part of an attack on the mining industry. In addition, the strict emission limits for new combustion plant in the LCP Directive were perceived as a threat to coal markets by the NCB itself,[27] giving the coal industry as a whole the sense that it was under attack from environmental demands.

The long-term policy which would reduce coal dependency was a major programme of PWR investment, agreed in October 1979. Although the primary justification for the programme was economic, as PWRs were at the time believed to be much cheaper than coal-fired stations, the security benefits for the electricity supply system in terms of fuel diversity and resilience to supply interruptions were also a major issue. The PWR policy was boosted with the appointment of Walter Marshall as Chairman of the CEGB in 1982.[28] Sir Walter, knighted shortly after his appointment, was a long-standing proponent of the Westinghouse PWR nuclear design and had previously been Chairman of the Atomic Energy Authority and Chief Scientist at the DEn.

The delicate task of presenting and interpreting evidence on projected energy demand at the Public Inquiry into the first PWR at Sizewell, and the importance of not publishing potentially conflicting views outside the Inquiry framework, placed some constraints on the government. In evaluating the possibility of joining the 30% Club, the government, 'deliberately', in the words of the DoE,[29] chose the DEn Sizewell energy scenario resulting in the highest projection of SO_2 emissions in order to calculate the amount of FGD retrofitting required. This resulted in an estimate of 8000 MW of FGD by 1993 at a cost, in 1984 money, of £600m, virtually guaranteeing Treasury opposition.

A lower estimate of unconstrained SO_2 emissions may have been more realistic. However, had the government argued that joining the 30% Club was easy, it might have cast doubts on the credibility of some of the scenarios which the DEn had submitted at the Sizewell Public Inquiry and might also have implied that coal markets were expected to decline in the 1990s.

The lack of any argument that nuclear power might make a contribution to reducing acid emissions, at Sizewell or elsewhere, is an indicator of the sensitivity of the government in 1983–5. The nuclear argument was deployed frequently in the FR Germany and has been used regularly in the UK since 1988 with reference to carbon dioxide emissions and the climate change problem.

High-level decisions

Issues such as acid-rain controls would normally be settled through the customary official and Ministerial Committee structure which operates within Whitehall. However, as in 1988 with the greenhouse effect, the Prime Minister took a personal interest, perhaps because the size and nature of the acid-rain problem appealed to her scientific training. It has been suggested that the Prime Minister enjoyed discussing such issues with CEGB Chairman Sir Walter Marshall because 'it flatters her self-image to sit down and have a scientific discussion with him'.[30]

At the end of May 1984, Mrs Thatcher called an informal weekend seminar on acid rain at the Prime Minister's country home, Chequers, in order to pull together the views of eminent scientists and representatives of the government departments directly concerned.[31] Ad hoc meetings such as these, outside the formal Whitehall Committee system, have been a common feature of the Thatcher administrations. The seminar was portrayed as a briefing for the Prime Minister but it effectively played a much more important role in fixing government policy at a critical time.

In June, an international conference on acid rain was to be held at Munich with Britain represented at the ministerial level. It was at this meeting that the Soviet Union and other Eastern bloc states backed the 30% Club. The position adopted by Britain at the conference would set the tone for negotiations on the LCP directive due to begin later in the month.

Attendees at the Chequers meeting included: representatives of the Treasury; the Head of the Meteorological Office, Sir John Mason; the DoE's Under Secretary of State, William Waldegrave, and its Chief Scientist, Martin Holdgate; CEGB Chairman Sir Walter Marshall and head of the Technology Planning and Research Division Peter Chester; and Sir Hermann Bondi, former DEn Chief Scientist and Chairman of the Natural Environment Research Council.

The Prime Minister herself was able to judge the arguments for and against tighter emission controls. The DoE did not ask for support for the draft LCP Directive, but are reported to have 'argued forcefully' that Britain should join the 30% Club.[32] In the event, the DoE lost its case and arguments advanced by the CEGB and others about the insufficient certainty of scientific evidence won the day. Britain went to the Munich conference with no specific emission reductions on offer.

The British debate about acid rain comprised two separate discourses, one scientific and one political. As part of the scientific discourse, the Chequers meeting had a curious mix of participants. Of the qualified scientists taking part, four were physicists or mathematicians (Marhsall, Mason, Bondi, Chester) while only one was a biologist (Holdgate). The CEGB's physicists certainly argued against emissions controls. Only a few days after the Chequers meeting, Martin Holdgate was to remark to a Parliamentary Committee that 'in so complex and multi-disciplinary an area as the natural environment, certainty is a luxury which we can very rarely enjoy. It may be more common with our mathematical and physical sciences . . . as to the political and policy side, I have to put on a different hat.'[33]

The presentation of the hard line

The Chequers meeting, which gave Britain's policy the seal of prime ministerial approval, set the policy tone for some time ahead. Later in 1984, a Cabinet

Committee was established under the Chairmanship of Lord Whitelaw, the deputy Prime Minister and the Conservative Party's leader in the House of Lords, to consider responses on acid rain.[34] That Mrs Thatcher herself did not chair the Committee, as she was to do with the Committee preparing the 1990 White Paper on environment policy, was a sign that acid rain was not regarded as a primary political issue.

In spite of the defeat at Chequers, the DoE twice pressed Cabinet to approve the retrofitting of FGD at a small number of power stations but were turned down.[35] However, the government's reply to the Environment Select Committee report, published in December 1984, contained the limited undertaking that SO_2 and NO_x emissions would each be cut by 30 per cent below 1980 levels by the end of the 1990s. This would have required little in the way of emission controls on existing plant.

As is inevitable under the British system of collective responsibility, it fell to the DoE which had argued for emission controls to explain to Parliament and the public why they were inappropriate. The CEGB's arguments about inconclusive evidence and excessive control costs were clothed in elegant terms:

> Pollution is dealt with by political action, but it is explained by science. Science is dynamic, and the policies of this and other Governments must evolve to meet new evidence about the environmental situation. What is durable within this framework of change is the Government's overall policy; that action against pollution shall rest on the best scientific evidence, the best technical and economic analysis, and the best possible assessment of priorities.[36]

Nevertheless, the government was not entirely sure of its parliamentary ground. The debate on the Environment Select Committee report and the government's reply in January 1985 was scheduled for a Friday afternoon, just as MPs customarily head back to their constituencies. Although the debate was, inevitably, sparsely attended, almost unanimous criticism of the government's policy came from both sides of the House. No vote was called and, in spite of a follow-up report from the Environment Committee in November 1985, the continuing efforts of environmental groups were insufficient to maintain parliamentary interest in acid rain as the issue dropped from the headlines.

Acid abatement proposals, 1986–87

New pressures

During 1984 and 1985, the European Commission attempted to isolate the UK during the Brussels negotiations on the LCP directive. The negotiations brought some sense of inevitability that, at some point in the future, the UK would need to make concessions. The UK held the Presidency of the European Council of Ministers during the second half of 1986 and it was necessary for the UK to make constructive proposals on the draft LCP Directive. Without some form of move on acid-emissions reduction, there was little prospect of doing so.

At the same time, the DoE continued to give thought to strategies for initiating an emissions abatement programme in the UK. In early 1986, an internal report produced by the Central Policy Studies Unit in the Department recommended that a pilot FGD plant should be constructed at one power station, with others to follow if initial operating experience was satisfactory.[37] In March 1986, the DoE announced that it would review acid-rain policy with the DEn and the CEGB in the light of recent developments in science, notably in Scandinavia.

Impressive results were beginning to emerge from experiments under way, notably the RAINS (Reversing Acidification in Norwegian Soils) project. This was beginning to demonstrate conclusively that the acidity of water entering lakes and streams could be substantially reduced, on a time scale of one to ten years, if deposition levels fell.[38]

The net result was that, in contrast to the situation in 1984, when the DoE was operating in a purely reactive mode, by 1986 it was actively trying to formulate positive policy steps within the very narrow bounds imposed by the outcome of the Chequers meeting two years earlier and the views of other interested departments.

The CEGB reconsiders

By 1986, with the Sizewell Inquiry and the miners' strike over, the CEGB began to play a less important role as an instrument of government policy. At the same time, the DoE/DEn/CEGB review of acid-rain policy made the CEGB aware of the continuing interest in emission controls within the DoE. In June 1986, Lord Marshall, the CEGB Chairman (ennobled in 1985), undertook a tour of Scandinavian research sites in the company of his environmental research director, Peter Chester, to see for himself some of the results of research.

It is evident that Lord Marshall was impressed by the conclusiveness of the experiments which he saw. On his return to the UK, a CEGB proposal was put together for the retrofitting of FGD at two UK power stations – Drax in Yorkshire, part of which had been commissioned earlier in 1986, and Fiddler's Ferry in Cheshire. The CEGB also proposed installing FGD at all new coal-fired power stations. The proposal was put on the DoE in July and gained rapid assent, with Treasury approval for the associated £600m capital expenditure programme.

The Industrial Air Pollution Inspectorate (IAPI), the body formally responsible for regulating emissions from UK power stations, was not consulted, leaving some uncertainty about the relationship between the CEGB initiative and the formal system of 'best practicable means' (BPM) controls exercised by the IAPI.

The decision was made public in September just as the Prime Minister was paying a visit to Oslo. In spite of the British gesture, Mrs Thatcher's visit was greeted by unprecedented scenes of violence in the Norwegian capital as demonstrations were held against UK environmental policy.

The FGD programme proposed was not urgent. The first half of Drax would be retrofitted by 1993, seven years after the decision, the second half in 1995 and Fiddler's Ferry by 1997. Thus a 6000 MW programme would take a decade to complete, constrasting with the 37 000 MW programme completed in five years in FR Germany.

The justification for the FGD programme was set out in some detail in a CEGB paper written in July 1986 in advance of the submission of the FGD proposal to the government.[39] This argued that the recent Scandinavian experiments had demonstrated that the acidity of water flowing into lakes and streams was determined primarily by the stock of sulphur chemically locked up in soil systems. The only way to reduce acidity was to deplete these sulphur stocks by reducing the quantities of sulphur being deposited. It was argued that a 'proportional' contribution to this effort on the part of the CEGB would involve it ensuring that 'taking one year with another SO_2 emissions will steadily fall between now and the end of the century'.

Hence, the modestly scaled and timed FGD programme would enable this objective to be met, smoothing out a projected rise in UK SO_2 emissions during the mid-1990s.

While the as-yet unpublished results of projects carried out under the CEGB/NCB's SWAP programme contributed to the CEGB's analysis, the single most influential piece of work was the Norwegian RAINS experiment. There was a view, however, in Scandinavia that this experiment did not so much uncover new facts as spell out in clear terms the implications of existing knowledge. Possible links between German forest damage and air pollution, about which the CEGB was profoundly sceptical, played no role in the policy re-assessment.

The implications of the FGD decision

The CEGB's FGD proposal was unilateral. It did not emerge through negotiations with the DoE. However unprompted the decision might have been, it proved enormously useful to the UK government during its Presidency of the Council of Ministers. As will be described more fully in Chapter 12, the UK was able to table a dignified, if implausible compromise proposal on the LCP Directive.

The FDG decision also helped the CEGB to retain the initiative in determining emission controls. In 1987, the CEGB was to announce the construction of two new coal-fired power stations to meet an expected shortage of power-plant capacity in the 1990s. By declaring its intention of installing FGD at any such stations, the CEGB remained one step of the IAPI. Later, in April 1987, the CEGB announced that it would retrofit low-NO_x burners to its largest coal-fired power stations over the period 1988–98 at a cost of £180m.

Agreement of emissions reductions, 1988

Electricity privatization proposals

As the LCP negotiations dragged on into 1987 and then 1988, major changes were in the offing for the electricity supply industry. In its 1987 General Election Manifesto, the Conservative Party announced that it would privatize the industry. Having viewed the privatization of British Gas en bloc in 1986 as something of a failure in terms of promoting competition and curbing monopoly powers, the government's supporters were keen to see a more radical approach. Some commentators viewed the break-up of the CEGB into as many as five or six regional utilities as a possibility.[40] During the latter half of 1987, the CEGB lobbied for its own preservation.

The government's White Paper, published in February 1988, came as a disappointment to the CEGB. The statutory obligation to supply electricity, which had underpinned the CEGB's primacy within the industry from 1957 onwards, was to be taken away and given to the electricity distributors and suppliers. In addition, it was planned (but not executed in this precise form) that the CEGB would be broken up into three parts: one company, to be owned jointly by the distribution companies through a holding company, with responsibility for transmission; one company with two-thirds of the generating capacity including all the nuclear plant; and one company with the remaining generation assets.

The broader implications of electricity privatization are discussed in Chapter 13. In the context of the ongoing discussions during 1987–8 about emission controls, the developments underline the rapidly widening gulf between the CEGB and the government. The CEGB could no longer count on government support when the most fundamental assumptions about its own structure and identity had been overturned.

In 1988, the question beginning to vex the City institutions charged with selling shares in the electricity supply industry was the attractiveness of the different components of the industry to potential investors. The largest questions were being raised about National Power, the inheritor of the CEGB's nuclear assets, because of uncertainties about the cost of decommissioning the ageing Magnox nuclear reactors expected to cease operating during the 1990s.

However, the market value of the industry would also be dependent on a wide range of factors concerning its regulation, the associated financial arrangements and the size of any future investment programmes. The question of future expenditure requirements on environmental controls, which might not necessarily be recovered through higher prices, was to have some bearing on the ultimate decisions on acid-rain policy.

The need for an abatement plan

Just after the publication of the privatisation White Paper, British negotiators found themselves isolated in discussions on the LCP Directive. There appeared to be no way forward in Europe unless some further concessions were made by the UK. It was the DoE's task to persuade the CEGB, as well as colleagues in the Treasury and the DEn, to consent to such a step.

By early 1988, the CEGB had carried out much more detailed engineering studies of the feasibility of installing FGD at individual power stations. In spite of the theoretical provision of space for retrofitting FGD in the planning consents for each of the major power stations, this space had subsequently been used for other purposes at some stations, occasionally without the knowledge of CEGB headquarters. By 1988, these constraints had been evaluated and views had been formed about the size, cost and timescales for different FGD programmes.

An important factor underlying the timescales was the insistence of the CEGB that power stations should not be taken off-line to hook up FGD units other than during scheduled maintenance times which take place every three years. This largely explains the relatively long time scale of the British FGD programme.

Britain's negotiating position was established by ministers the day before the make-or-break meeting of the Council of Ministers on the LCP Directive. In contrast to May 1984 when the CEGB had been able to present its case directly to the Prime Minister, it was not directly involved in this policy decision.

The factors influencing the final policy stance must remain a matter of speculation. However, relevant considerations are likely to have included the size of the feasible FGD programmes suggested by the CEGB and the DoE's assessment of what would be acceptable to the EC Commission and the German Presidency of the Council of the Ministers. Also taken into account would have been the impact of FGD retrofitting costs on electricity privatization prospects, the effects on competition within a privatized electricity supply industry and, conversely, the degree of uncertainty which would have been introduced by any failure to reach an agreement. The possibility of even higher environmental demands being made if a decision were delayed may also have been a consideration.

Agreement is reached

In June 1988, the UK agreed to cut its emission of SO_2 from existing combustion plant by 20 per cent, 40 per cent and 60 per cent below the 1980 level by the years 1993, 1998 and 2003 respectively. In practice, the CEGB believed that this would mean the installation of two additional FGD retrofits by 1998 and one more by 2003. The DoE assumed that three additional FGD retrofits would be necessary by 1998 with a further one by 2003.[41]

This was not as much as the European Commission and other EC Member States had been seeking. However, it had extracted from the UK, after five years of negotiating, just over half of the SO_2 abatement from existing power stations which might have been required under the 1983 draft of the LCP Directive.

Implementation proposals, 1989–90

The problems of implementation

Following formal agreement of the LCP Directive in November 1988, consideration began to be given to the implementation arrangements in the UK. The government faced three inter-related challenges in complying with its EC obligations:

1   to bring a new primary legislation in order to place UK implementation of the 1984 'framework' Directive on industrial emissions on a firm legal footing;
2   to make specific plans for allocating emission reduction requirements for existing plant to specific sectors, companies or sites within the framework of altered primary legislation; and
3   to ensure consistency with the new arrangements for a competitive market in bulk electricity supply which would follow privatization of the power industry.

This chapter focuses on the UK's emissions reduction plan and the narrower question of consistency with privatization arrangements. Primary legislation and privatization arrangements in general are described in Chapter 13.

Implementation philosophies

The UK implementation of the LCP Directive is underpinned by three different pollution control philosophies, all embodied in the Environmental Protection Act 1990. This did not receive the Royal Assent by 30 June 1990, the date by which the Directive requires Member States to 'bring into force the laws, regulations and administrative provisions necessary for them to comply'. The control principles are:

1   uniform standard setting for new installations;
2   a limited market incentive based approach for existing plant through a system of company emission totals; and
3   technology specification by a regulator exercising administrative discretion as embodied in the BPM approach. In practice, this element of control may be overshadowed by the first two.

Whether tensions emerge between the different control philosophies will depend greatly on the extent to which environmental problems relating to acid emissions in the UK emerge as an issue and the degree to which further European pressures are brought to bear.

The emission limits for new plant in the LCP Directive are to be directly translated into British law using powers created under the Environmental Protection Act. Implementing the requirements to abate emissions from existing plant has proved much more complex. Here, the UK was required to draw up a programme for the progressive reduction of emissions from existing plant by 1 July 1990 and to inform the Commission of this programme by the end of 1990.

In drawing up an implementation plan, two fundamentally different approaches were possible. The first was to make specific abatement requirements at individual plants, while the second was to allocate emission totals to individual companies, leaving them to decide which of their plants to modify and what types of abatement measures to apply. A third option, adopted in the FR Germany, that of establishing uniform emission standards for every plant was not considered as this would have implied emission reductions (and costs) far in excess of the LCP Directive requirements.

The plant-specific approach, recognizing that the degree of environmental damage depends on the emissions location, would allow the development of a plan which would maximize the environmental benefits at sites sensitive to acid deposition within the UK. Work carried out by the Atomic Energy Research Establishment at Harwell under contract to the DoE has shown that the maximum environmental benefit would be realized by fitting FGD at power stations unlikely to be favoured by the power industry on cost grounds alone.[42] The plant-specific approach would almost certainly be necessary if the concept of 'critical loads' for sensitive ecosystems being developed within the UN Economic Commission for Europe (and supported by the UK) were to be applied in practice.

However, it was the company/industry emissions total approach which was adopted and underlay the DoE's August 1989 consultation paper outlining how the government proposed to implement the requirements of the LCP Directive.[43]

The first step towards compliance will involve the DoE establishing a statutory national plan under Section 3(5) of the Environmental Protection Act specifying permitted emissions of SO_2 and NO_x by year and industrial sector. Companies operating several LCP may be allocated company emission totals. These will certainly include the electricity generators but may also include major industrial companies.

The second step will involve HMIP and the other Inspectorates granting authorizations for individual LCP in a manner consistent with: (1) the Secretary of State's national plan; (2) use of best available techniques not entailing excessive cost (BATNEEC); (3) Community Treaties or other international law; and (4) air quality standards or objectives. For multi-site companies given emission totals under the DoE's national plan, the Inspectorates will discuss the allocation of permitted emissions among individual authorizations.

For enforcement purposes, emission quotas will be set one year ahead. When emissions reach a 'reporting level' below that quota, plant operators will be obliged to inform the Inspectorate. Quotas for further years ahead will have the status of planning totals. Operators awarded a company quota may approach HMIP to have the quotas allocated to individual plants altered as long as the total company allocation is not breached. If a multi-site company given an emission total wishes to change that figure, it must approach the DoE in order to have the statutory national plan altered.

The allocation of company totals embodies, in a very limited sense, the

Table 11.2   LCP Directive: indicative possible sectoral breakdowns of emission reductions (ktonnes SO^2)

Sector	1980	1987	% red[1]	1993	% red[2]	1998	% red[2]	2003	% red[2]
Power stations	3007	2830	6	2691	11	1804	40	1203	60
					(5)		(36)		(57)
of which – National									
Power	1654	n.a.	n.a.	1497	9	984	40	653	60
PowerGen	1125	n.a.	n.a.	1019	9	670	40	444	60
Industry	600	307	49	205	66	123	70	92	85
					(33)		(60)		(70)
Refineries	282	162	43	141	50	121	57	101	64
					(13)		(25)		(38)
Total	3889	3299	15	3038	22	2048	47	1396	64
					(8)		(38)		(58)

Notes:
(1) % reductions from a 1980 base;
(2) figures in brackets from a 1987 base.

Sources: Department of the Environment, National Power, PowerGen.

concept of emissions trading since companies will be able to transfer emission 'rights' from one site to another as long as the company total is not breached. This shifts discretion from the regulator to industry which is, in principle, in a better position to assess its own abatement costs and draw up implementation plans which meet a given emissions target at the lowest possible cost. It is possible that a demand for wider emissions trading may arise as pairs or groups of companies identify mutual advantages in adjusting their emissions totals.

While the operational responsibility for implementation of the Directive at the plant level will rest with HMIP, the application of the company total concept is likely to reduce its authority to specify emission reductions at any particular plant, making it difficult to take into account local or even regional environmental factors.

The government's emissions reduction plan

The DoE's August 1989 Consultation Paper contained an 'indicative possible sectoral breakdown' of SO_2 emissions reductions in line with the EC timetable. A second Consultation Paper containing a more detailed plan was promised. The final decisions about how much abatement to seek from each individual sector of the LCP stock will be based on considerations of 'practicality, equity and economics'.

The 'indicative scenario' for reducing SO_2 emissions from existing plant, shown in Table 11.2, has proved one of the more controversial aspects of the first part of the DoE's consultation exercise. The most striking feature is the small contribution of the electricity supply industry to the emissions reductions required by 1993. A 5 per cent reduction from a 1987 base compares with a proposed 33 per cent reduction for industry and 13 per cent for petroleum refineries. The low contribution of the electricity supply industry is attributed to the long lead times associated with the FGD retrofit programme. For the second- and third-stage emission reduction targets, the distribution of SO_2 abatement is more evenly spread round the different sectors. However, even so, industry is expected to reduce emissions by 70 per cent between 1987 and the year 2003 compared to 57 per cent for the power industry.

Electricity supply arrangements

The CEGB could, as a state-owned industry, simply fit certain power stations with FGD and run them on base load in order to keep within an SO_2 ceiling. This could be done by formal government requirement, through BPM specification by HM Inspectorate of Pollution (HMIP) or through some less formal understanding analogous to the coal-supply arrangement with British Coal.[44]

However, integrating SO_2 control into the post-privatization electricity supply arrangements is necessarily more complex. The new arrangements for electricity supply require individual power stations to compete with each other to supply a 'power pool'. Because stations fitted with FGD have higher running costs, there is no guarantee that they will be able to compete successfully against uncontrolled coal-fired plant. With unconstrained competition, compliance with an SO_2 ceiling might not be possible even if plants were fitted with FGD.

The problem was acknowledged explicitly by the CEGB when it announced the awarding of a contract for its first FGD retrofit, Drax, in February 1989. Lord Marshall floated the idea of a 'smoke tax'[45] which would be levied on power stations *without* FGD, improving the competitive position of FGD stations and providing funds to subsidize the capital costs of FGD construction. The 'smoke tax' would have led to SO_2 abatement being financed through increases in electricity prices, estimated at 2 per cent.

However, during the course of 1989, two substantial revisions were made to the power pooling arrangements. The setting of formal contracts between electricity generators and distributors, with the cost of FGD built in, was abandoned. In addition, options for SO_2 abatement other than FGD began to look more attractive.

The solution which has emerged is simpler from the regulatory point of view but more complex for the electricity generators. In October 1989, a single generators/distributors power pool was agreed which is essentially a spot market for electricity operating on a half-hour basis. There is no 'smoke tax', and the generators are free, as long as they do not abuse their market power, to bid whatever price for the output of their power stations that they wish.

As a result, consumers will not bear the cost of the emissions reduction programme because electricity prices in the power pool will be determined by uncontrolled power stations. The polluter pays principle will not therefore apply as far as the emission reduction programme for existing power stations is concerned. National Power and PowerGen will need to operate power stations fitted with FGD in order to stay within the company SO_2 totals set by the DoE. To do so, they may need to bid prices below the actual cost of operation.

These arrangements will reduce the profitability of stations fitted with FGD and lower their worth, reducing the asset value of National Power and PowerGen and hence Treasury receipts from privatization. Another effect is to shift the incentives for different types of SO_2 abatement. A 'smoke tax' would effectively have been a subsidy for FGD and, therefore, indirectly for higher sulphur British coal. Without a 'smoke tax', the use of lower sulphur imported coal has become relatively more attractive.

The electricity supply industry's abatement plans

When the LCP Directive was being negotiated, and for some time afterwards, both the CEGB and the government believed that FGD would be the principle means used to achieve SO_2 abatement. The DoE's August 1989 Consultation Paper spoke of industry plans for a 12 000 MW FGD programme, a view re-iterated by the Prime Minister in a speech to the UN General Assembly in November of that year.

However, in early 1990, it became clear that the electricity generators had identified a far wider range of options for abating SO_2, including increased use of low sulphur imported coal and the rapid introduction of combined cycle gas turbine stations in order to replace old coal-fired plant. By that time, contracts had been placed for only 4000 MW of FGD at Drax. In early 1990, PowerGen raised the possibility that they would undertake no FGD retrofits at all. If only Drax were retrofitted with FGD, British Coal anticipated a market for its production of only 20 m tonnes by the year 2003 compared to 75 m tonnes in 1990–1. The size of Britain's FGD programme became a matter for an extensive debate between the interested parties, much of it conducted in the national press.

The change in abatement strategy is explained partly by the removal of fundamental constraints on the electricity generators' operations. Privatization gave National Power and PowerGen the opportunity to use imported coal in quantities established soley by commercial considerations. At the same time, the European Commission had given an indication that gas use in new, highly efficient combined cycle gas turbine plant would not be constrained by the 1975 Directive restricting gas use in power stations.[46] On purely economic grounds, increased use of low sulphur imported coal and gas were increasingly attractive.

During the first half of 1990, the electricity generators were negotiating with the DEn about their future financial structure, with the DoE about the size of their company SO_2 totals and with British Coal about the details of their coal contracts over the period 1990–3. They were also positioning themselves for subsequent negotiations about contracts beyond 1993.

The generators used uncertainty about the size of the FGD programme as a lever in negotiations with British Coal in order to obtain price concessions. Given the LCP Directive, large quantities of high sulphur British coal could be used only if power stations were fitted with FGD. In order to make this cost-effective for the generators, National Power argued that the delivered price of British coal would have to be £5–6/tonne below that of imported coal.[47]

There were rumours of a disagreement between the DEn, which wished to leave the technical measures for SO_2 abatement to the discretion of the electricity generators, and the DoE, which had put the case to other EC Member States in June 1988 that the UK could not accept lower SO_2 ceilings because of the excessive costs of a larger FGD programme.[48] The DoE felt that its European position had been undercut by the electricity industry's change of tack on SO_2 control.

In the event, PowerGen had decided, in the light of its SO_2 reduction requirements, to build 4000 MW of FGD capacity at Ferrybridge and Ratcliffe power stations, matching National Power's programme. The present decision to build a total of 8000 MW of FGD does not necessarily mean that further FGD will not be built at a later stage. The generators have emphasized that no decisions have yet been made about how their year 2003 SO_2 emission reduction targets might be met. According to the DoE, the planning period for a firm emissions reduction plan must correspond to the longest lead time for all the technical abatement options.[49] This will be FGD which, in the UK, might take five-six years to plan and construct.

In the meantime, it is unlikely that FGD will be fitted to new coal-fired power stations in the UK. Most of Britain's needs for new capacity during the 1990s will be met by combined cycle gas turbine plant. Beyond the year 2000, to the extent that coal-fired plant is installed it will employ advanced combustion technologies such as PFBC or integrated gasification combined cycle (IGCC), giving rise to SO_2 and NO_x emissions which are intrinsically lower.

In April 1990, the House of Commons Energy Committee conducted a short Inquiry into Britain's FGD programme.[50] The Committee concluded that the government had 'not yet faced up to the dilemma which confronts it over the future of Britain's coal industry'. The possibility that the European Commission might, during its 1994 review, seek to increase the UK's emission abatement requirements for the year 2003 in light of the scaled-down FGD programme and broader questions about comparability of effort between the UK and other Member States was another issue which engaged the Committee. This Inquiry made it clear that, in spite of the signing of the LCP Directive, the debate about UK policy on acid emissions is unlikely to disappear. The challenges for the 1990s are discussed in Chapter 13.

Notes

1. Department of the Environment (1972), *Pollution: Nuisance or Nemesis?*, HMSO, London.
2. Sir Hugh Rossi, House of Commons Energy Committee (July 1984), *Acid Rain*, 4th Report, Session 1983–4, HC 446, HMSO, London, Question 641.
3. Environmental Data Services Ltd (November 1985), *ENDS Report 130*, p. 5.
4. The most recent of the Review Group reports are: UK Review Group of Acid Rain, 1987, *Acid Deposition in the UK 1981–1985*, Second report, Warren Spring Laboratory, Stevenage; UK Terrestrial Effects Review Group, 1989, *The Effects of Acid Deposition on the Terrestrial Environment in the UK*, HMSO, London; UK Acid

Waters Review Group, 1989, *Acidity in UK Fresh Waters*, Second report, HMSO, London; and Building Effects Review Group, 1989, *The Effects of Acid Deposition on Buildings and Building Materials in the UK*, HMSO, London.

5. Dr Peter Chester, Director, Technology Planning and Research Division, CEGB, House of Commons Energy Committee (July 1984), *Acid Rain*, 4th Report, Session 1983–4, HC 446, HMSO, London, Question 69.

6. National Coal Board, memorandum in House of Lords Select Committee on the European Communities (June 1984), *Air Pollution*, 22nd Report, Session 1983–4, HL 265, HMSO, London, para. 17.

7. Central Electricity Generating Board (1984), *Annual Report and Accounts 1983/84*, London, para. 11.

8. Commission of the European Communities, 15 December 1983, *Proposal for a Directive on the Limitation of Emission into the Air from Large Combustion Plants*, COM (83)704 final, Brussels.

9. Environmental Resources Limited (1983), *Acid Rain: a Review of the Phenomenon in the EEC and Europe*, EUR 8684, Graham and Trotman for the Commission of the European Communities, London.

10. Nature Conservancy Council (1984), *Acid Deposition and Its Implications for Nature Conservation in Britain*, Focus on nature conservation No 7, London; and N. Dudley *et al*, (1985), *The Acid Rain Controversy*, Earth Resources Research, London.

11. C. Rose and M. Neville (1985), *Tree Dieback Survey: Final Report*, Friends of the Earth, London.

12. Watt Committee on Energy (1984), *Acid Rain*, Report Number 14, London.

13. For example: S. Elsworth (1984), *Acid Rain*, Pluto Press, London; F. Pearce (1987), *Acid Rain*, Penguin, Harmondsworth; J. McCormick, *Acid Earth: The Global Threat of Acid Pollution*, Earthscan, London.

14. CBI evidence, House of Commons Energy Committee (July 1984), *Acid Rain*, 4th Report, Session 1983–4, HC 446, HMSO, London, p. 191.

15. House of Lords Select Committee on the European Communities (June 1984), *Air Pollution*, 22nd Report, Session 1983–4, HL 265, HMSO, London.

16. House of Commons Energy Committee (July 1984), *Acid Rain*, 4th Report, Session 1983–4, HC 446, HMSO, London.

17. T. Paterson (June 1984), *A Role for Britain in the Acid Rainstorm*, Bow Publications Ltd, London.

18. Centre for Policy Studies (1985), *Greening the Tories: New Policies on the Environment*, London.

19. *Inspecting Industry: Pollution and Safety (1986)*, Efficiency Scrutiny Report, HMSO, London.

20. Central Electrcity Generating Board, Memorandum in House of Lords Select Committee on the European Communities (June 1984), *Air Pollution*, 22nd Report, Session 1983–4, HL 265, HMSO, London, p. 200.

21. Environmental Data Services Ltd (June 1984), *ENDS Report 113*, p. 3.

22. P. Hennessy (1986), *Cabinet*, Basil Blackwell, Oxford, p. 32.

23. Ibid.

24. Central Electricity Generating Board (1982), *Annual Report and Accounts 1981–82*, London.

25. J. Winterton and R. Winterton (1989), *Coal, Crisis and Conflict: The 1984–85 Miners' Strike in Yorkshire*, Manchester University Press, Manchester.

26. Kim Howells MP, former National Union of Mineworkers researcher, on *Newsnight*, BBC TV (22 May 1990).

27. National Coal Board evidence, House of Commons Energy Committee (July 1984), *Acid Rain*, 4th Report, Session 1983–4, HC 446, HMSO, London, p. 279.

28. 28 November 1987, The current favourite, *The Spectator*, pp. 19–20.

29. Department of the Environment evidence, House of Commons Energy Committee (July 1984), *Acid Rain*, 4th Report, Session 1983–4, HC 446, HMSO, London, p. 61.
30. 28 November 1987, The current favourite, *The Spectator*, pp. 19–20.
31. Environmental Data Services Ltd (May 1984), *ENDS Report 112*, London, p. 3.
32. Environmental Data Services Ltd (June 1984), *ENDS Report 113*, London, p. 3.
33. Department of the Environment (December 1984), *Acid Rain: The Government's Reply to the Fourth Report from the Environment Committee, Cmnd 9397, HMSO, London, question 172.*
34. *P. Hennessy (1986), Cabinet, Basil Blackwell, Oxford.*
35. Memorandum submitted by Greenpeace, in House of Commons Energy Committee, June 1990, *The Flue Gas Desulphurisation Programme*, Third Report, Session 1989–90, HC 371, HMSO, London, p. 41.
36. Department of the Environment (December 1984), *Acid Rain: The Government's Reply to the Fourth Report from the Environment Committee*, Cmnd 9397, HMSO, London, para. 1.4.
37. Environmental Data Services Ltd (February 1986), *ENDS Report 133*, London, pp. 3–4.
38. Dr Ivar Muniz, House of Commons Environment Committee, *Air Pollution*, First report, Session 1987–8, HC 270, HMSO, London, Question 156.
39. P. Chester (1986), *Acid Lakes in Scandinavia: The Evolution of Understanding*, CEGB, London.
40. A. Holmes *et al.* (1987), *Power on the Market*, Financial Times Business Information, London.
41. Environmental Data Services Ltd (June 1988), *ENDS Report 161*, London p. 3.
42. R. Derwent (1988), *Optimization by Simulated Annealing and an Optimal Strategy for Retrofit Flue Gas Desulphurisation in the UK*, AERE-R13110, HMSO, London.
43. Department of the Environment (August 1989), *Implementation of the Large Combustion Plants Directive: A Consultation Paper*, London.
44. The National Coal Board was renamed the British Coal Corporation in 1986.
45. Financial Times Business Information, 2 March 1989, UK smoke tax – the power industry cleans up, *Power in Europe*, No. 14, pp. 7–8.
46. House of Commons Energy Committee (June 1990), *The Flue Gas Desulphurisation Programme*, Third report, Session 1989–90, HC 371, HMSO, London, Question 88.
47. Ibid, Question 12–14, p. 62.
48. Acid rain clean-up 'cut-back' (17 February 1990), *The Independent*, p. 1.
49. House of Commons Energy Committee, op. cit., question 177.
50. House of Commons Energy Committee, op. cit.

12 Compromise on acid rain: the role of the European Community

Introduction

While the UN Economic Commission for Europe (UNECE) and the machinery of the Long Range Transboundary Air Pollution (LRTAP) Convention promoted acid emission abatement at the wider European level, the European Community (EC) played a decisive role in securing more ambitious policies in its twelve Member States. However, the political desire to reduce emissions within the EC was far from uniform. The FR Germany sought the transfer of its domestic policies to other Member States, although this happened only in a considerably weakened form. In doing so, it had the support of other environmentally ambitious countries, notable Denmark and the Netherlands. The UK led the considerable resistance with tacit support from the less industrialized members of the Community.

This chapter tells how one of the EC's principal measures dealing with acid emissions from power stations and other stationary plant, the Large Combustion Plant (LCP) Directive, was initiated and negotiated. British – German tensions are a major theme in this story. The negotiation of vehicle emissions regulations, which followed a similarly complex course, is described briefly in Chapter 13.

To place the LCP Directive in context, the EC's legislative competence in the environmental field and its first three Action Programmes on the environment are first described. The description is not comprehensive as there are already a number of excellent descriptions of EC environmental legislation and policy.[1]

Prior to describing the LCP Directive, the preceding 'framework' Directive on air pollution from industrial plant is discussed. The 'framework' Directive established the EC's competence to take detailed measures to curb atmospheric emissions through 'daughter' Directives of which the LCP Directive is, so far, the only example.

The implications of the Single European Act and other more recent developments for EC environment policy are then described. The chapter concludes with a short discussion of British and German approaches to the LCP negotiations.

The legal framework for EC environmental policy

The EC is unique as a multinational organization in that it has supra-national powers, i.e. it has acquired a degree of sovereignty ceded by Member States.

One of the main motivations of the EC founder members was the creation of a 'common market' among Western European nations. Consequently, a great deal of EC activity has focused on the removal of barriers to trade.

The Single European Act has made significant changes to the legal basis for Community action in the environmental field. However, both the industrial 'framework' Directive and the LCP Directive were effectively developed under the pre-1987 rules which are described in this section.

Prior to 1987, most environmental measures were justified with reference to Article 100 of the Treaty which empowers the Council of Ministers to 'issue directives . . . as directly affect the establishment or functioning of the common market' or Article 235 which allows that 'if action by the Community should prove necessary . . . and this treaty has not provided the necessary powers, the Council shall . . . take the appropriate measures'.

EC policy is implemented through the issuing of Regulations, Decisions and Directives. Regulations automatically become directly binding law in all Member States. Decisions are binding on those to whom they are addressed. Directives however have been the main tool used to implement EC environmental policy. Only the objectives of these are binding on Member States, which have discretion to determine implementation at the national level.

Directives must be proposed by the European Commission, effectively the EC's civil service. These must be approved unanimously by the Council of Ministers which is the ultimate decision-making body. However, the role of the Council is often more pro-active than this procedure would suggest, as it frequently invites the Commission to put forward proposals.

The Council of Ministers formally consists of the foreign ministers of each Member State. However, foreign ministers are generally represented by ministers from the appropriate departments of national government when specific topics are under discussion. Thus, environmental policy decisions are taken by the 'Council of Environment Ministers'. The Environment Council meets between two and four times a year.

The Presidency of the Council is rotated round Member States at six month intervals. For the President of the Council, there are considerable political incentives to secure unanimous agreement and maximise the amount of legislation passed. According to one participant 'it is the test of the success of a Presidency'.[2]

The European Parliament and the EC's Economic and Social Committee must be consulted about proposed Directives which would require the amendment of legislation in any Member State.

The action programmes on the environment

The first action programme, 1973–76

The EC entered the field of environmental protection by adopting a coherent policy at the time of the 1972 Stockholm Conference on the Human Environment, just before Britain, Denmark and Ireland joined the six founder members. A 'Programme of Action Concerning the Environment', covering the period 1973–6 was agreed by the Council of Ministers in 1973.[3] Three other Action Programmes have followed. The main objective of the First Action Programme was to:

bring expansion into the service of mankind by procuring for mankind an environment providing the best conditions of life and to reconcile this expansion with the increasingly imperative necessity of preserving the natural environment.

The principles to be applied included:

1   the prevention of pollution at source rather than subsequently trying to counteract the effects;
2   the polluter pays principle;
3   that activities carried out in one state should not degrade the environment in another; and
4   the harmonization of Community and national environment policies in a common long-term plan.

In practical terms, the First Action Programme on the Environment produced the Gas Oil Directive which restricted the sulphur content of gas oil in Member States.[4] A sizeable research programme was also instituted as part of the First Action Programme.

The second action programme, 1977–81

During the Second Action Programme,[5] a Directive on Ambient Air Quality was issued, laying down limit values for concentrations of SO_2 and suspended particulates.[6] The final compliance date for this Directive is 1993.

In its review of the Second Action Programme, the EC Commission noted that lack of resources, lack of political will, differences in the responsiveness of the Member States, scientific uncertainties and enforcement problems had all hindered progress with formulating a Community environmental policy.

The third action programme, 1982–86

The Third Action Programme, established just as acid rain was emerging as a major political issue,[7] was philosophically more ambitious than its two predecessors. For the first time, the Commission established a number of priorities for action, noting that:

1   'it will continue its efforts to establish air quality standards'; and
2   'for some ubiquitous pollutants a policy will have to be devised which will stabilise (standstill principle), and thereafter gradually reduce (rollback principle), total emissions by establishing emission standards, where necessary for certain pollutants'.

The Commission also introduced a number of new principles which would finally re-emerge as part of the Single European Act, including the need to integrate environmental questions into other policies and the need to take a preventive as opposed to a reactive approach to environmental problems.

In March 1985, ambient air quality standards were extended to cover NO_x.[8] The final compliance date for this measure is 1 January 1994, although Member States were required to 'take the necessary measures to ensure' earlier compliance by 1 July 1987. The Third Action Programme also saw the adoption of the 'framework' Directive on industrial air pollution and the Commission's proposal for the LCP Directive which are described in more detail below.

The framework directive on air pollution from industrial plants

The first concrete step towards an EC agreement on a reduction in acid emissions came in June 1982 when, at the time of the Stockholm Conference on the acidification of the environment, the FR Germany both joined the '30% Club' and sent the EC Council of Environment Ministers a request that priority should be given to developing a Community framework for the prevention of air pollution. After two Environment Council discussions, the Commission issued a draft 'Directive on combatting of air pollution from industrial plants' in April 1983.[9]

Discussions on this draft Directive lasted just over a year before it was agreed, in a slightly modified form, in March 1984. It was signed in June 1984 after the lifting of UK Parliamentary reserve.[10] The Commission's proposal had gained authority after the discussion of environmental issues for the first time at the European Council meeting in June 1983. Forest dieback in Germany had also been discussed at this meeting and it is clear that German concerns were very much driving the European debate.

The UK participated in the unanimous decision of the Council of Environment Ministers to adopt this directive without being in full accord with all of the provisions.[11] The UK was unhappy about the principle of EC powers to set Community-wide emission limits. However, the original Commission proposal had foreseen emission limits being agreed by a qualified majority in the Council of Ministers and the UK succeeded in having this changed to unanimity, thus retaining the option of exercising a veto on any future proposal. Ever cost-conscious, the UK also succeeded in having other key changes made – the proposed requirement to use 'best available technology' at sites requiring prior authorization was qualified with the phrase 'not entailing excessive costs' and the compliance date was delayed from 1 January 1985 to 1 July 1987.

The formal objective of the 'framework' directive is 'to provide for further measures and procedures designed to prevent or reduce air pollution from industrial plants'. It does not itself specify emission limits for any particular class of plant, but lays out the circumstances under which certain types of industrial plant may be licensed to operate and which plants and substances would be subject to controls. Plants in the energy sector covered by the Directive include: coke ovens; oil refineries; coal gasification and liquefaction plants; and thermal power stations (excluding nuclear power stations) and other combustion installations with a nominal heat output of more than 50 megawatts (MW). Polluting substances include SO_2 and other sulphur compounds, NO_x and other nitrogen compounds, and six other classes of pollutants.

The framework Directive provides:

1   that all listed industrial plants should require prior authorization before being allowed to operate;
2   that this authorization will also be required when plants are substantially altered;
3   that an authorization should be given only if (a) 'all preventive measures against air pollution have been taken, including the application of the best available technology, provided that the application of the such measures does not entail excessive costs'; (b) 'the use of plant will not cause significant air pollution'; (c) 'none of the emission limit values applicable will be exceeded'; and (d) 'all the air quality limit values applicable will be taken into account';
4   that 'the Council, acting unanimously on a proposal from the Commission shall if necessary fix emission limit values based on the best available technology not entailing excessive costs, and takin into account the nature, quantities and harmfulness of the emissions concerned'; and
5   that 'Member States shall implement policies and strategies . . . for the gradual adaptation of existing plant . . . to the best available technology, taking into account in particular the plant's technical characteristics, its rate of utilisation and length of its remaining life, the nature and volume of polluting emissions from it, and the desirability of not entailing excessive costs for the plant concerned, having regard in particular to the economic situation of undertakings belonging to the category in question'.

The effective implementation date for the directive is 1 July 1987, as existing plants are defined as those built or authorized before that date.

The large combustion plant draft directive

The LCP Directive is the first and, so far, the only daughter Directive deriving from the 1984 framework Directive. There were four broad phases in its long and complex development: the origins of the Directive in Federal German legislation in 1982–3, early fruitless negotiations covering the period 1984–5; the initiatives of the Council Presidencies in 1986–7; and the final stages 1987–8.

The origins of the directive

The LCP Directive is modelled very much on German domestic legislation.[12] According to the UK Department of the Environment (DoE), the European Commission's LCP proposal and the 'framework' Directive were conceived simultaneously in 1982.[13] At that time, German legislation had been drafted, but was not to be enacted until June 1983. In a sense, it was the perceived need for the very specific LCP Directive which stimulated the need for the broader 'framework' legislation. The Commission released the draft LCP Directive in December 1983.

In the explanatory memorandum justifying its proposal,[14] the European Commission cited the 1982 German memorandum to the Council of Ministers on air pollution, the 1982 Stockholm Conference on the acidification of the environment, the calls at the June 1983 European Council meeting for drastic air pollution control measures as a matter of priority, and the 'conclusions widely accepted' at the September 1983 Karlsruhe symposium, 'Acid Rain: A Challenge for Europe'. According to the Commission, these were that there was a 'need to bring about a considerable reduction as soon as possible in the volume of SO_2 and NO_x emissions'.

A further motivation for the Commission was to take a step towards harmonizing national legislation and to remove 'the unequal conditions of competition' created 'by differing levels of protection against air pollution'. The 'equalization of conditions of competition' argument could be advanced because the cost implications of the domestic German legislation of June 1983.

The Commission's justification for the LCP Directive was widely challenged in the UK. Several bodies, including the National Society for Clean Air and the Institute for European Environmental Policy questioned the Commission's interpretation of the 'widely accepted' scientific conclusions of the Karlsruhe symposium,[15] which the Commission had itself sponsored, and the interpretation of the scientific evidence on acid rain.[16] There was a widespread suspicion in the UK energy industries that the Commission's proposal had been unduly influenced by domestic German pressures. The EC's rapid change of position (it had been instrumental in opposing Scandinavian proposals for the 30% Club at the Executive Board meeting of the LRTAP Convention earlier in 1983) was also a source of resentment.

The Commission was also critized for basing its proposals on economic factors as opposed to environmental need, underestimating the costs of compliance and misrepresenting the state of science as regard acid rain.[17] In particular, it is alleged that the proposal was based on a single consultant's report which had not been subject to review.[18]

The timescale for preparation of the draft Directive was also a concern. The Commission's internal consultations, and those with Member States and interested parties, which normally precede a complex EC Directive, may take years to complete. For the LCP Directive, the time-scale was a few months. National governments, industrial interests and even sections of the European Commission felt excluded from a hasty and ill-considered drafting process. The Environment Directorate's haste can, in retrospect, be seen as a mistake. When negotiations on the draft Directive began in 1984 a great deal of resistance, much of it British, emerged.

The·Commission's 1983 draft

The December 1983 draft of the LCP Directive had two distinct objectives:

1   the setting of Community-wide limits for emissions of SO_2, NO_x and
     particulate matter from new large combustion plant, i.e. that with an output
     of more than 50 MW thermal authorized after 1 January 1985; and
2   reductions in total emissions of SO_2, NO_x and particulate matter from all
     LCP in each Member State by 1995 calculated from a 1980 base level.
     The national emissions 'bubbles' were based on reductions of 60 per cent
     for SO_2, 40 per cent for NO_x and 40 per cent for particulate matter.

Tables 12.1–12.5 compare the original proposals with the significantly
modified provisions of the Directive as finally negotiated in June 1988. Broadly
speaking, for new plants the draft Directive required flue gas desulphurization

Table 12.1   Sulphur dioxide emission limits for new plants – LCP Directive
(milligrammes per normal cubic metre[1])

Fuel and Capacity[2]	Final Agreement	Commission Proposal 1983
Solid Fuels[3]		
50–100 MW	[4]	2000
100 MW	2000	2000
100–500 MW	2000–400[5]	2000/400[6]
500 MW+	400	400
Liquid fuels		
50–300 MW	1700	1700
300–500 MW	1700–400[5]	400
500 MW+	400	400
Gaseous fuels		
as a rule	35	35
liquefied gas	5	5
coke oven gas	800	100

Notes:
(1) measured on a dry basis on 0°C, atmospheric pressure and 6% oxygen for solid fuels, 3%
    oxygen otherwise;
(2) based on thermal input;
(3) for plants burning indigenous solid fuel where the emission limits would entail excessively
    expensive technology, desulphurization rates apply instead. These are: 40% for plant sized
    100–167 MW; 90% for plant above 500 MW; and a sliding scale between 40% and 90% for
    intermediate plant;
(4) to be decided by the Council in 1990;
(5) on a sliding scale based on plant size;
(6) 2000 milligrammes for plant below 300 MW or grate firing in plant of any size.

Source: European Commission.

Table 12.2 Nitrogen Oxide Emission Limits for New Plants – LCP Directive (milligrammes per normal cubic metre[1])

Fuel and capacity[2]	Final Agreement	Commission Proposal 1983
Solid fuels		
as a rule	650	800
with ‹10% volatiles	1300	–
pulverized firing with extraction of fused ash	–	1300
Liquid fuels	450	450
Gaseous fuels	350	350

Notes:
(1) measured on a dry basis at 0°C, atmospheric pressure and 6% oxygen for solid fuels, 3% oxygen otherwise;
(2) based on thermal input.

Source: European Commission.

Table 12.3 Dust Emission Limit for New Plants – LCP Directive (milligrammes per normal cubic metre[1])

Fuel and capacity[2]	Final Agreement	Commission Proposal 1983
Solid Fuels		
500 MW+	50	50
‹500 MW	100	50
Liquid Fuels	50[3]	50
Gaseous Fuels		
as a rule	5	5
blast furnace gas	10	10
gases produced by the steel industry used elsewhere	50	100

Notes:
(1) measured on a dry basis at 0°C, atmospheric pressure and 6% oxygen for solid fuels, 3% oxygen otherwise;
(2) based on thermal input;
(3) 100 milligrammes for plant below 500 MW burning fuel with an ash content greater than 0.06%.

Source: European Commission.

Table 12.4 Nitrogen oxide emission reduction targets for existing plant[1]

	1980 emissions (ktonnes)	1987 adjusted emissions	1993 reduction (%)	1998 reduction (%)
Belgium	110	110	20	40
Denmark	124	135	3	35
FR Germany	870	870	20	40
Greece	36	70	(94)	(94)
Spain	366	460	(1)	24
France	400	400	20	40
Ireland	28	50	(79)	(79)
Italy	580	713	2	26
Luxembourg	3	3	20	40
Netherlands	122	122	20	40
Portugal	23	64	(157)	(178)
UK	1016	1016	15	30
EC 12	3678	4013	10	30

Note:
(1) The original Commission proposal was for a uniform 40% reduction by 1995.

Source: European Commission

(FGD) plant, or clean coal technology, for all installations sized above 300 MW thermal and low sulphur fuel for smaller installations. It was proposed that the SO_2 emission limits should be tightened after 1995. NO_x emission limits implied the use of low-NO_x burners, but not more expensive selective catalytic reduction. For existing plants, the emission ceilings implied either retrofitting or premature closure of power stations and larger industrial installations on a major scale.

Early negotiations, 1984–85

The initial reception for the draft LCP Directive in early 1984 was mixed. While FR Germany, the Netherlands and Denmark were behind the proposal, and France and Belgium could afford to be indifferent because of their nuclear power programmes, the proposal presented major problems for the UK, Italy, Greece and Ireland.

The UK quickly emerged as the most committed opponent of the proposal, claiming that the evidence linking acid emissions to environmental damage was inconclusive, that the clean-up bill was unacceptably high and that the 1980 starting point for emissions reductions was unfair since UK emissions had already declined substantially during the 1970s.

Conversely, from the Commission's point of view, it was particularly important that the UK play a full part in any agreement. The UK was a major industrial economy, the largest SO_2 emitter in the Community and a major contributor to acid-rain damage. The Commission's central objectives of reduced acid emissions and harmonization of environmental controls would be thwarted without the UK's full participation.

While the UK led the opposition to the draft Directive, the smaller Member States with reservations sheltered behind the UK position. Consequently, the Commission's strategy during 1984–5 was to isolate the UK from its less vocal supporters. This was achieved by offering specially tailored concessions ('derogations') to the smaller countries – Greece, Ireland and Luxembourg. However, by the end of 1985 no agreement had been reached.

In January 1986, Portugal and Spain become EC members. Spain in particular faced major difficulties with the LCP Directive because of a rapid growth in coal consumption for power generation. This factor complicated the negotiating position and doomed to failure the Commission's early strategy of attempting to isolate the UK.

The role of the Council Presidencies, 1986–87

From 1986 onwards, the initiative in negotiations passed from the Commission to those Member States which, in turn, held the Council Presidency. The largest issue facing negotiators was that of emission reduction targets for existing plant, particularly those for SO_2. Table 12.5 shows the proposed SO_2 emission reductions for each Member State which were tabled over the period 1986–8. During that time, significant modifications were also made to the emission limits proposed for new plant. Both issues are discussed below.

THE DUTCH PRESIDENCY – EARLY 1986

The Netherlands held the Council Presidency for the first half of 1986 and introduced a proposal whereby the proposed SO_2 reduction would take place over two stages, the first stage securing a 45 per cent reduction for the EC as a whole by 1995 and the second securing a 60 per cent reduction by the year 2005. While the issue of the allocation of the 60 per cent reduction among Member States by 2005 was not addressed, an ingenious allocation of the 45 per cent reduction by 1995, which offered concessions to everyone, was drawn up.

A set of 'objective' criteria were used to identify how much each Member State should contribute. These included: contribution to total emissions; use of thermal power plant per capita; GDP per capita; trade in transboundary pollution; the state of economic development; and the use of problematic indigenous fuels. These were compared with the emission reductions expected through the continuation of current policies which would lead to an expected overall 30 per cent drop in emissions. Some Member States (Belgium, FR Germany, Denmark, France, the Netherlands) would over-achieve in terms of their 'objective' contribution to the 45 per cent reduction. These were not required to increase their efforts. Member States with a low per capita energy use were asked to have a standstill on SO_2 emissions (Greece, Ireland, Luxembourg, Portugal) or a 10 per cent reduction (Spain). The two remaining

Table 12.5
Proposed Sulphur Dioxide Emission Reductions 1983–88
Large Combustion Plant (LCP) Negotiations (ktonnes SO_2 and % reductions)

Proposer Proposal date Target date	Total SO_2 1980	LCP SO_2 1980	Adjusted LCP 1987	CEC 1983 1995	NL 1986(i) 1995	 2005	UK 1986(ii) 1995	 2005
Belgium	799	530	530	60	50	*	57	57
Denmark	437	323	355	60	50	*	67	51
FR Germany	3200	2225	2225	60	70	*	58	58
Greece	400	303	582	60	0	*	0	0
Spain	3250	2290	2899	n.a.	10	*	6	31
France	3560	1910	1910	60	60	*	91	91
Ireland	219	99	175	60	0	*	0	43
Italy	3800	2450	3000	60	40	*	47	62
Luxembourg	23	3	3	60	0	*	0	0
Netherlands	467	299	299	60	60	*	77	77
Portugal	328	115	309	n.a.	0	*	0	0
UK	4670	3883	3883	60	40	*	28	52
EC 12	21153	14430	16169	60	45	60	41	54

Notes:
(1) UK proposals were for total emissions, not just those from LCP;
(2) * = to be decided at a later date;
(3) 'proposers' were: CEC = European Commission; NL = Netherlands; B = Belgium; UK; DK = Denmark; FRG = FR Germany;
(4) n.a. = not applicable;
(5) figures in brackets represent permitted emissions increases.

	B 1987(i)			DK 1987(ii)			FRG 1988(i)			*Directive* June 1988		
*	1993	1998	2005	1993	1998	2010	1993	1998	2003	1993	1998	2003
*	40	60	60	40	60	80	40	60	70	40	60	70
*	34	56	56	34	56	78	34	56	67	34	56	67
*	40	60	60	40	60	80	40	60	70	40	60	70
*	3	3	3	(15)	(15)	23	6	6	21	(6)	(6)	(6)
*	24	37	37	16	30	62	0	24	37	0	24	37
*	40	60	60	40	60	80	40	60	70	40	60	70
*	(25)	(25)	(25)	(41)	(41)	29	(25)	(25)	(10)	(25)	(25)	(25)
*	26	51	51	27	42	76	27	39	63	27	39	63
*	40	50	50	40	50	60	40	50	60	40	50	60
*	40	60	60	40	60	80	40	60	70	40	60	70
*	(28)	(28)	(28)	(107)	(129)	(8)	(29)	(42)	(26)	(102)	(135)	(179)
*	26	46	60	22	33	80	26	46	70	20	40	60
60	31	49	53	27	42	74	25	43	60	23	42	57

countries (Italy, the UK) were asked to reduce their emissions further (a 40 per cent reduction) by 1995.

This proposal failed because some countries, notably Ireland and Spain, found even a standstill in SO_2 emissions difficult. To that extent, an important function of the Dutch proposal was that it flushed out the views of those countries which had hidden behind the UK position. At the same time, the Germans and the Commission objected to a 'three-speed' Europe. Because of these other objections, the UK had no need even to respond to the suggestion that it should cut SO_2 emissions by 40 per cent by 1995.

THE UK PRESIDENCY – LATE 1986

The UK had the Presidency for the second half of 1986 and used the Dutch proposals as the starting point for a new compromise. Potential UK embarrassment was relieved somewhat by the CEGB's volunteering to retrofit FGD to 6000 MW of power plant in July 1986.

The UK proposed a 30 per cent EC-wide reduction in SO_2 by 1995 and a 45 per cent reduction by 2005. A 60 per cent reduction would be allocated at an indeterminate point in the future. Introducing a new element into the negotiations, the UK proposed that these reductions should be from all sources, not just LCP. (This was attractive to the UK because its own SO_2 emissions from small sources had declined considerably since 1980 because of increased natural gas use, industrial decline and energy conservation.)

The UK proposal abandoned the Dutch concept of objective criteria and was based, rather optimistically, on political acceptability. As Table 12.5 shows, the proposal contained huge discrepancies in the abatement efforts required from different Member States. Those countries which expected to achieve large emission reductions in any case were asked to do even more. Thus France, with a large nuclear programme, was asked to achieve a 91 per cent cut in SO_2 emissions between 1980 and 1995.

This proposal forced some countries to make their positions even clearer. Spain and Ireland announced, for the first time, that they could not accept even the principle of either emission limits for new plant or limits on their total SO_2 emissions. Both expected their emissions to rise substantially above 1980 levels and were not prepared to install expensive control equipment because of the impacts on their industrialization.

The UK also made suggestions on emission limits for new plants, in particular suggesting no emission limits for plant sized below 100 MW and a sliding scale for emission limits based on plant size. This was based on a domestic proposal by the UK Industrial Air Pollution Inspectorate issued in mid-1986. The rationale was to avoid giving artificial incentives to operators to size plant just below the 300 MW threshold previously proposed. There was a wider acceptance of this positive aspect of the UK proposals. The sliding scale proposal, though not the 100 MW cut-off point, was to become a feature of the Directive as finally agreed.

THE BELGIAN PRESIDENCY – EARLY 1987

In early 1987, Belgium brought the smaller countries back into the negotiations and introduced those elements which were to lead to a successful conclusion. Spain and Ireland agreed, with the other Member States, to 'contribute' to a solution to the LCP Directive logjam.

The Belgians went back to uniform emission reduction levels, but in three stages (1993, 1998 and 2005). However, emission reductions were then modified to take account of two factors.

The first was the change in LCP capacity installed in each country between 1980 and 1987. This addressed the issue of rapidly increasing levels of coal use in some countries. The second factor was the change in emission levels prior to 1980, addressing the UK's major concern. All Member States agreed with the two principles introduced by the Belgians.

Ironically, the starting point of uniform emission reduction levels from an altered baseline acted as a device to gain final acceptance for the principle of final reduction levels which were *not* uniform. Thus, a major point of difference between the UK, the smaller countries, the more environmentally ambitious Member States and the Commission, was removed.

The Belgian proposal completed the process of flushing out the views of the smaller and newer EC members. This revealed that the final Directive would need to contain derogations which met the specific circumstances of the less industrialized Member States. However, the Belgian proposal did not satisfy all of the Spanish concerns.

Belgium kept the UK concept of a sliding scale of emission limits for new plants, but re-introduced limits for coal plants sized between 50 and 100 MW. Apart from this factor and some adjustments to the NO_x emissions limits, the Belgian proposal for new plant was very similar to that which was finally agreed.

Nevertheless, in spite of a highly constructive initiative, the Belgian Presidency had still failed to produce a consensus.

THE COMMISSION'S ROLE

Compared to 1984–5, the Commission played a relatively minor role during the Dutch, UK and Belgian Presidencies. The Netherlands and the UK did not consult the Commission about their proposals. However, the Dutch idea of 'objective' criteria to allocate emissions reductions found some sympathy with the Commission, though not with other Member States.

A major concern for the Commission during the Belgian Presidency was the degree to which its original proposals had been watered down so as to be barely acceptable. There was some confidence that negotiations on new plant emission limits could be concluded, but also a recognition that a major political compromise was needed for existing plant.

The Commission's only real power was that of veto. Two options were possible: (1) to withdraw the original proposal entirely and start from scratch; or (2) to split the Directive into two parts, one covering new plant emission limits and the other existing plant emission ceilings. Both options risked major delays. In particular, splitting proposals on new and existing plants might have rendered an agreement on emission reduction targets impossible to attain. This aspect of the negotiations was vital if the Community was to be able to sign the Helsinki 30% Protocol on SO_2 emissions.

In the event, with the two succeeding presidencies belonging to the environmentally more ambitious Danes and Germans, the Commission did not take any drastic steps and negotiations continued.

The conclusion of negotiations, 1987–88

THE DANISH PRESIDENCY – LATE 1987

The Danish Presidency's proposal in late 1987, built on the Belgian initiative, was received broadly favourably by those countries fighting large SO_2 reductions. The first- and second-stage SO_2 reduction targets for Greece, Spain, Ireland, Portugal and the UK were considerably less ambitious than those proposed by Belgium. Denmark also postponed the third stage for reductions to the year 2010, but suggested an EC-wide SO_2 reduction of 75 per cent by that time. Many countries were unsatisfied by both the ambitiousness and the remoteness of this target. This reduced, however, the apparent contentiousness of the targets for the years 1993 and 1998.

Denmark acceded to the UK's request for a 100 MW threshold for the regulation of SO_2 emissions from coal-fired plant. It also retained, again at the UK's suggestion, a sliding scale for SO_2 removal efficiencies as an alternative to emission limits for new plant.

While the more ambitious Member States were concerned that Britain was getting off the hook in the Danish proposals, there was no comparable concern about the loosening of emission reduction requirements for other countries. In the end, Denmark did not push its proposals hard, partly because of the potential difficulty in obtaining approval from its own Parliament.

By the end of 1987, the political will among all parties to complete the negotiations had hardened and a considerable degree of optimism had emerged. The FR Germany, the 'father' of the proposed Directive, was keen to complete the process it had begun five years before and bring an agreement back home. The fact that subsequent Council Presidencies belonged to Greece and Spain, neither of which was likely to press for a solution, added further urgency to this desire. The Germans signalled at the December 1987 Environment Council that they were likely to toughen up the Danish proposals.

THE GERMAN PRESIDENCY – EARLY 1988

The main issue to be determined during the German Presidency was the question of SO_2 emissions reductions. As shown in Table 12.5, starting from the Danish position, the Germans increased the UK's SO_2 emission reduction targets while reducing those for Spain and Italy. The magnitude of the Danish third stage emission reduction target was scaled back, but advanced to the year 2003. The Germans also proposed NO_x reductions of 25 per cent by 1993 and 40 per cent by 1998. In addition, thorny questions of a more technical nature remained to be settled, including the 50 or 100 MW cut-off point for regulation of new coal-fired plants, the averaging time for emission measurements and the application of emission limits to plants being extended.

At an informal Council meeting held at the Bavarian resort of Wildbad Kreuth in February 1988, the Germans pushed very hard and succeeded in completely isolating the UK, even on the narrower technical questions. From this point onwards, the UK's relations with the Environment Council and the domestic political agenda, with electricity privatization under way, were completely enmeshed. A strong political desire both to avoid European isolation and to clarify emission abatement requirements produced a broader flexibility and a willingness to compromise.

However, the March Environment Council proved a near disaster for the LCP negotiations. After an apparently constructive bilateral meeting on the eve of the Council, the Germans pushed the UK very hard to accept a 50 MW threshold for coal-fired plants, but were rebuffed by the British negotiating team which employed arguments about the impacts on domestic coal production. Feeling at the meeting was reportedly very anti-British and the Germans terminated discussions, without addressing the larger question of SO_2 emissions reductions, after only a short period.

Subsequently, the German Presidency began a series of bilateral meetings with the UK and Spain in particular. The Commission played a considerable role at this point in trying to broker a solution, siding with the UK on issues such as monitoring requirements and the treatment of plant extensions. Some of these disputed issues were settled before the June Environment Council.

There was huge pressure on all sides to produce a result at the Environment Council of 16 June. There was also a feeling that only Germany had the authority to deliver a compromise faced with historic British resistance. The meeting lasted until 4 a.m. on the 17 June, but did, in the end, deliver a heavily modified LCP Directive.

The provisions of the LCP Directive

Tables 12.1–12.5 contain the provisions of the LCP Directive. The elements of the final compromise were:

1   SO_2 reduction targets for the UK of 20 per cent by 1993, 40 per cent by 1998 and 60 per cent by 2003.
2   basic reductions of NO_x set at 20 per cent for 1993 and 40 per cent for 1998, modified for the circumstances of individual Member States;
3   the 50 MW threshold to stay, but emission limits to be fixed by the Council in 1990 after a Commission report on the availability of low sulphur fuel;
4   a monthly averaging time for emission measurements; and
5   plant extensions to be treated as separate installations for the purpose of setting emission limits;

The Directive also contained derogations which met the needs of individual Member States:

1   new plants larger than 400 MW operating for less than 2200 hours per year
    are allowed to operate at twice the otherwise permitted level of SO_2
    emissions (800mg/Nm^3);
2   new plants burning indigenous solid fuel which cannot meet the SO_2
    emissions limits without the use of excessively expensive technology may
    exceed the limits as long as specified rates of desulphurization are achieved
    (see Table 12.1);
3   Spain is allowed to construct certain amounts of new coal-fired electricity
    generating capacity with relaxed controls up to the year 2000;[19] and
4   new plants burning indigenous lignite may exceed emission limits if there
    are major difficulties connected with the lignite and it is an essential source
    of fuel.

In addition, 'if a substantial and unexpected change in energy demand or in the
availability of certain fuels or certain generating installations creates serious
technical difficulties for the implementation by a Member State ... the
Commission may at the request of the Member State concerned ... take a
decision to modify ... the emission ceilings'. However, any Member State can
refer such a decision to the Council which may make a different decision on the
basis of qualified majority.

In 1994, the Commission is required, 'where necessary', to propose a
revision of the year 2003 ceilings for SO_2 and the 1998 ceilings for NO_x. The
Commission must also submit proposals for the revision of the new plant
emission limits by July 1995.

Much of the text of the LCP Directive is devoted to matters such as
monitoring requirements, provisions of multi-fuel firing and provisions covering
the contingencies of the breakdown of abatement equipment or the non-
availability of low-sulphur fuels. In contrast to the situation in the US, plants
may, within limits, continue to operate with abatement equipment out of
service.

Although the Directive was agreed in June 1988, it was not formally signed
until November 1988 because of the time taken by national translators and
jurists to agree final texts.[20] The SO_2 emission reductions secured by the
Directive will not be sufficient to allow the EC to become a signatory to the
Helsinki Protocol to the LRTAP Convention.

Future combustion plant controls

There is a possibility that the European Commission may introduce a Small
Combustion Plant Directive covering plant down to 5 or 10 MW in size. This
has been the Commission's ambition since the early 1980s, in order to prevent
the anticipated displacement of higher sulphur fuels into smaller plant as a
result of agreement on the LCP Directive. However, given the fierce arguments
over emission limits for plant sized between 50 and 100 MW, which are still not
resolved, it may be difficult to secure any agreement with teeth.

Developments in EC environmental policy

The Single European Act – the environmental title

Since the LCP Directive was conceived, the Single European Act[21] has given EC environmental policy a firm legal footing by adding Title VII, which specifically addresses environmental issues, to the Treaty of Rome. However, certain environmental measures may be agreed under more general provisions of the Treaty, facing lower political hurdles. Formally, the LCP Directive was agreed under Title VII.

Article 130r of the amended Treaty of Rome sets out the objectives and principles underlying environmental policy. The objectives are:

1 'to preserve, protect and improve the quality of the environment';
2 'to contribute towards protecting human health'; and
3 'to ensure a prudent and rational utilisation of natural resources'.

However, the Member States annexed a Declaration to the Single European Act to the effect that the EC's environmental activities should not interfere with national energy policies.

The principles guiding EC action are that:

1 'preventive action should be taken';
2 'environmental damage should as a priority be rectified at source'; and
3 'the polluter should pay'.

The factors to be taken into account in preparing action are:

1 'available scientific and technical data';
2 'environmental conditions in the various regions of the Community';
3 'the potential benefits and costs of action or of lack of action'; and
4 the economic and social development of the Community as a whole and the balanced development of its regions'.

In addition, 'environmental protection requirements shall be a component of the Community's other policies'. Article 130t of the revised Treaty allows Member States to adopt protective measures which are more stringent than the Community norm. Voting at Council meetings on proposals made under Article 130 must be unanimous. However, the Council may define certain matters on which decisions may be taken by a qualified majority. This concept is defined below.

Environmental measures and the internal market

Certain environmental Directives which have as their object 'the establishment and functioning of the internal market' may be agreed under Article 100a of the Treaty of Rome which allocates less power to individual Member States and more to the EC's own institutions. In these cases, the Council may agree by

Table 12.6 Distribution of national votes in the EC Council of Ministers

Belgium	5	Ireland	3
Denmark	3	Italy	10
FR Germany	10	Luxembourg	2
Greece	5	Netherlands	5
Spain	8	Portugal	5
France	10	UK	10

Note: 54 votes out of the total of 76 are needed for a qualified majority.

qualified majority rather than unanimously. For a qualified majority to be established, 54 of the 76 votes available must be cast in favour of a Directive. The distribution of votes round Member States is shown in Table 12.6. A measure can be opposed only if at least three Member States object, and two of these would have to be larger countries such as Germany, France, Italy or the UK. If the Council wishes to amend a Commission proposal, it must do so unanimously.

If a measure has been adopted under Article 100a by qualified majority voting, Member States are allowed to make separate national provisions in order to protect the environment. This provision, included in the Single European Act at the insistence of the Danish Government, is not attractive to the European Commission.

Under Article 100a, the Council must work in cooperation with the Parliament rather than simply consult it. The cooperative procedure is somewhat complex. In essence, Parliament may, by an absolute majority, either reject or propose amendments to a Council position. If Parliament rejects a proposal, the Council must then act unanimously for a Directive to be agreed. If Parliament agrees amendments, the Commission must then re-examine its proposal. If the Commission accepts the Parliament's amendments then the Council may confirm it by a qualified majority. However, if the Commission rejects Parliament's amendments, of if the Council wishes to amend a proposal re-examined by the Commission it must act unanimously.

Product standards, such as fuel quality or motor vehicle emission standards, are the types of measures most likely to be covered by Article 100a. Environmental measures such as species conservation would certainly have to be implemented using Article 130r requiring a unanimous decision by the Council. Emission limits for stationary plant lie somewhere between these two examples. In principle, it is possible that the Commission might use Article 100a, which presents a lower political hurdle.[22] However, the 'framework' Directive on industrial air pollution specifically requires daughter Directives to be determined unanimously. Further, the Commission's legal service is expected to discourage the use of Article 100a for Directives with primarily environmental objectives.

The Fourth Action Programme, 1987–92

The EC's Fourth Action Programme for the Environment,[23] agreed in 1987, is even more ambitious than the first three. It points out the 'challenge to make a definite move away from reacting to environmental problems after they have arisen towards a general preventive approach . . . and at the same time to find means of deriving economic and employment gains through such a move'. This echoes a German rather than a British approach to environmental policy.

In drafting the Fourth Action Programme, the Environment Directorate sought to link its plans to those for completing the internal market in 1992. It placed emphasis on the need to harmonize environmental standards in order to prevent the erection of new barriers to trade. However, given that the Environment Directorate has relatively small resources, the full implementation of its proposals is very unlikely.

There is a particular emphasis on air pollution in the Fourth Action Programme and, specifically, the Commission intends to propose emission limits for all the major sectors listed in the 'framework' Industrial Plant Directive, define targets for acid-emission reductions and consider the setting of emission limits for smaller combustion sources, air-quality standards for photochemical oxidants and 'ecological air quality standards' for sensitive areas. How much of this will actually be achieved depends on how the new Articles in the Treaty of Rome are used and on the degree to which the Commission and the environmentally ambitious Member States can persuade the newer and less industrialized countries to participate in implementing expensive pollution controls.

Negotiating styles and lessons

While the LCP negotiations appear to have evolved into an intergovernmental haggle which might have taken place in any international forum, with the European Commission playing almost a background role, the fact that the discussions took place within the EC framework had an important bearing on the outcome. The final compromise was more than a 'lowest common denominator' solution, since countries such as the UK and Spain were pushed much further in the direction of emissions control than they had originally wanted to go. This outcome was assisted by the fact that the participants were operating within a larger web of obligations and commitments entailed by EC membership.

The path to the LCP Directive was complicated by the very different styles and expectations of the negotiating partners, notably the UK and the FR Germany. While the views of participants are necessarily subjective, a number of distinct themes emerge.[24]

The vulnerability of the German position stemmed from its insistence on translating domestic legislation to the Community level. This led to a great deal of inflexibility and insufficient sensitivity to the wider European context. German negotiators tended to have a good technical grasp of the subject and the basic premise underlying many of their arguments was that they were essentially 'right'. This led to a degree of rigidity on what appeared to others to be minor technical matters, even at Environmental Council level. Officials of the European Commission now regret having acted so 'obviously' in taking domestic German legislation, in great technical detail, as the basis for European law.

The British negotiators, on the other hand, were not technically trained and tended to view discussions as a diplomatic game with the same elements as any other negotiating exercise. For them, technical matters defined the face value of the cards with which they obliged to play. There was less sense of the intrinsic merits of the arguments in their negotiating briefs. The British thus tended to take a more sophisticated view of the negotiating process itself.

However, the prime factor affecting the British position was, until the last minute, the very restrictive brief granted by the government in London. Not without reason have insiders likened the British environment negotiators during the 1980s to a soccer team with eleven very good goalkeepers. Standard diplomatic ploys, such as keeping fruitless discussion going until the early hours of the morning, were used constantly. Inevitably, this bred frustration and resentment on the part of negotiating partners who commented unfavourably on the technical grasp of the British team.

Thus, the difficulties of the LCP negotiations stemmed not only from fundamentally opposed interests but also from a superimposed clash of styles. The lesson learnt from the LCP years, 1983–8, should prove useful as the EC begins to address the even tougher question of a European climate change policy.

Notes

1. N. Haigh (1984), *EEC Environmental Policy and Britain: An Essay and a Handbook*, Environmental Data Services Ltd, London; N. Haigh (1987), *EEC Environmental Policy and Britain* (2nd edn), Longman, Harlow; S. P. Johnson and G. Corcelle (1989), *The Environmental Policy of the European Communities*, Graham & Trotman, London.
2. William Waldegrave, Minister of State at the Department of the Environment in oral evidence reported in Select Committee on the European Communities (May 1987), *Fourth Environmental Action Programme*, 8th Report, Session 1986–7, HMSO, London, p. 90.
3. Commission of the European Communities, Declaration on the Programme of Action of the European Communities on the Environment (20 March 1973), *Official Journal of the European Communities*, C112, Brussels.
4. Commission of the European Communities (27 November 1975), Directive on the approximation of the laws of the Member States relating to the sulphur content of various liquid fuels, 75/716/EEC, *Official Journal of the European Communities*, L307, Brussels.
5. Commission of the European Communities (13 June 1977), Resolution on the continuation and implementation of a European Community policy and action programme on the environment, *Official Journal of the European Communities*, C139, Brussels.
6. Commission of the European Communities (30 August 1980), Directive on air quality limit values and guide values for sulphur dioxide and suspended particulates, (80/779/EEC), *Official Journal of the European Communities*, L229, Brussels.
7. Commission of the European Communities (17 February 1983), Resolution on the

continuation and implementation of a European Community policy and action programme on the environment, *Official Journal of the European Communities*, C46, Brussels.

8. Commission of the European Communities (27 March 1985), Directive on air quality standards for nitrogen dioxide, (85/203/EEC), *Official Journal of the European Communities*, L87, Brussels.

9. Commission of the European Communities (April 1983), *Proposal for a Directive on the Combatting of Air Pollution from Industrial Plants*, COM(83) 173, Brussels.

10. Commission of the European Communities (16 July 1984), Directive on the combatting of air pollution from industrial plants, 84/360/EEC, *Official Journal of the European Communities*, L188, Brussels.

11. Environmental Data Services Ltd (March 1984), *ENDS Report 110*, London, p. 21.

12. Grossfeurungsanlagenverordnung (GFAVo) (July 1983), *13th Bundesimmission-schutzverordnung*, Bonn.

13. Department of the Environment in House of Commons Environment Committee (July 1984), *Acid Rain*, Fourth report, Session 1983–84, HMSO, London, Vol. 2, p. 76.

14. Commission of the European Communities (15 December 1983), *Proposal for a Directive on the Limitation of Emissions into the Air from Large Combustion Plants*, COM (83)704 final, Brussels.

15. Evidence reported in House of Lords Select Committee on the European Communities (June 1984), *Air Pollution*, 22nd report, Session 1983–4, HL265, HMSO, London, p. 95. and question 177.

16. Environmental Resources Limited (1983), *Acid Rain: A review of the Phenomenon in the EEC and Europe*, Graham & Trotman for the Commission of the European Communities, London.

17. House of Lords Select Committee on the European Communities, *Air Pollution*, 22nd Report, Session 1983–4, pp. xxiv–xxix, HL 265, HMSO, London, June 1984.

18. William Waldegrave, Minister of State at the DoE, in oral evidence reported in House of Lords Select Committee on the European Communities, *Fourth Environmental Action Programme*, 8th Report, Session 1986–7, p. 86, HL 135, HMSO, London, May 1987.

19. The Spanish derogations apply to plant larger than 500 MW. Up to 2000 MW of electric generating capacity burning imported solid fuel may be installed as long as it meets an SO_2 limit of 800 mg/Nm^3. A 60 per cent desulphurization rate is permitted for up to 7500 MW of electrical generating capacity, or 50 per cent of all new solid fuel capacity needs up to the year 2000 (whichever is the less), burning imported solid fuel.

20. Commission of the European Communities (7 December 1988), Council directive on the limitation of certain pollutants into the air from large combustion plants (88/609/EEC), *Official Journal of the European Communities*, L336, Brussels.

21. Commission of the European Communities (1986), Single European Act, *Bulletin of the European Communities*, 2/86, Brussels.

22. This issue was discussed in Environmental Data Services Ltd (February 1987), *ENDS Report 145*, pp. 9–11.

23. Commission of the European Communities (7 December 1987), Resolution on the continuation and implementation of a European Community policy and action programme on the environment, *Official Journal of the European Communities*, C328, Brussels.

24. The discussion is based on interviews with officials in the UK Department of the Environment, the European Commission and the German Federal Environment Ministry.

PART IV
Emerging issues and conclusions

13 Atmospheric pollution and the energy industries: from acid rain to climate change

Introduction

By 1988, a milestone in the regulation of acid emissions had been reached in Europe with the agreement of the European Community's (EC's) Large Combustion Plant (LCP) Directive and the attaching of the Sofia Protocol requiring a freeze on nitrogen oxides (NO_x) emissions to the Long Range Transboundary Air Pollution (LRTAP) Convention. With political commitments made, the focus began to move towards implementation, regulatory instruments and enforcement, particularly in the UK. However, in the FR Germany, the implementation of controls on acid emissions from stationary sources is virtually complete.

Nevertheless, the pace of change in environmental policy has quickened since 1988. There is room for the further development of acid-rain policy, since the European Commission must reconsider both plant-specific emission limits and overall emission reduction targets in 1994–5 and the Sofia Protocol requires signatories to negotiate NO_x emission reductions based on critical pollution loads.

At the same time, the environmental debate has widened to cover the issue of responses to possible climate change. This poses potentially even greater challenges to the energy industries than acid rain because technical fixes such as flue gas desulphurization (FGD) are not available and changes with major implications for the energy supply industries may be necessary.

There are other challenges. The opening up of Eastern Europe has begun to reveal the huge potential for emissions reductions from obsolescent industrial plant. The prime concern of Scandinavian governments regarding transboundary air pollution is now Poland rather than the UK. The unification of Germany will ensure that these issues will become a very direct concern for the EC.

Also, the EC's objective of completing the internal market by 1992 has implications for environmental policy as it relates to the energy sector. The Single European Act has put EC environmental policy on a firm legal footing and the 1992 drive towards liberalization of energy markets may help to change the traditionally close relations between governments and the electricity and gas utilities in particular. Given this changing context, it is possible that environmental controls could represent one of the few remaining levers offering government influence over the energy sector.

It is the purpose of this chapter to describe the state of the energy-environment debate as it stands in mid-1990, two years after EC agreement on acid rain, but on the threshold of even greater challenges. The issues discussed are: the nature of the policy problems posed by global environmental change; policies on atmospheric pollution at the European level, covering both the EC and UN Economic Commission for Europe (UNECE) regimes; and developments in Britain and Germany respectively. The country sections cover evolving acid rain policy, environmental policy in general, institutional and legal changes, developments within the energy sector and responses to the climate change issue.

The global environment

Global environmental issues have crept up the political agenda since the middle of the 1980s. In 1985, the Vienna Convention on the Protection of the Ozone Layer was agreed, though no binding commitments to reduce chlorofluorocarbons (CFCs) was accepted at that time.

Greater concern about global environmental issues was stimulated in 1987 when the UN's Brundtland Commission published its report,[1] giving wide currency to the concept of sustainable development and pointing out the complex interwining of economic and environmental problems. 1987 also saw the signing of the Montreal Protocol on CFC production. While this measure was not directed at energy production and use, it was the first example of a truly global measure aimed at restricting atmospheric emissions.

The significance of the Montreal Protocol for the energy industries is that it points the way towards the type of global agreement which will be necessary if steps are to be taken to mitigate emissions of carbon dioxide (CO_2) and other greenhouse gases. The majority of CO_2 emissions originate in the energy sector. Policies to address the climate change problem have become much debated in Britain, Germany and the EC as well as at the global level.

The potential for global warming was recognized within the scientific community long before it emerged as a political issue in the late 1980s. In 1979, the First World Climate Conference had discussed the possibilities of both global warming and a descent into a new ice age. However, by the time of the 1985 Villach Conference, co-organized by the World Meterological Organisation (WMO) and the UN Environment Programme (UNEP), a scientific consensus had emerged that global warming was a real possibility. In 1987, UNEP endorsed the idea, put forward at the Tenth World Meteorological Congress, of an intergovernmental mechanism to study all aspects of the climate change problem.

The Intergovernmental Panel on Climate Change (IPCC) was established in November 1988. The following month, the UN General Assembly formally requested the IPCC to deliver its report, including the identification of possible elements of a Climate Change Convention, within two years. UNEP is now geared up to begin negotiations following the delivery of the IPCC report to the UN in late 1990.

The 1988 Toronto Conference organized by the Canadian Government was also very important in creating a wider interest in the climate change issue. This conference called for a 20 per cent reduction in the CO_2 emissions of industrialized countries by the year 2005 starting from a 1988 base, followed by more drastic emissions reductions over a longer timescale. However, although governments sent delegates to the Toronto Conference, it was not a formal negotiating forum and the call for a 20 per cent emissions reduction had no status in international law. Rather, it acted as a political stimulus for governments to raise the priority attached to the climate change issue.

In order to address the enormously complex problem of climate change, the IPCC has been structured into three working groups:

1   *Science* – (chaired by the UK) covering the modelling of climatic changes, projection of possible futures and identification of gaps in knowledge;
2   *Impacts* – (chaired by the Soviet Union) covering the agricultural, economic and social impacts of climate change; and
3   *Responses* – (chaired by the US) covering the forecasting of future greenhouse gas emissions, control strategies and identification of the possible components of a Convention.

The first of the three reports to appear, that of Working Group 1, was unambiguous in the view that climate change is under way.[2] Within a considerable band of uncertainty, its central view was that global temperatures would rise by 1.1°C by the year 2030 and by 3.3°C by the year 2090 on the assumption of 'business-as-usual' greenhouse gas emissions. Sea level rise was estimated to be between 9 and 29 cm by the year 2030 and 28 and 96 cm by 2090. To stabilize atmospheric concentrations of CO_2, the Working Group estimated that emissions would have to be reduced by between 60 and 80 per cent.

The deliberations of Working Group III, particularly as regards energy/CO_2 emission projections and reduction targets, have been more controversial and of a more political nature. Some governments, notably the UK, felt that the global energy models being used to assist Working Group III needed to be supplemented by country case studies which would provide more realistic assessments of future greenhouse gas emissions. However, the officially prepared country case studies have tended to incorporate high emissions projections which suggest that cuts in emissions, or even stabilization, might be difficult to achieve.

Non-government organizations and independent analysts have taken a more optimistic view of future emission levels, presuming these to be lower and hence making emission reduction targets seem more easily achievable. Like the energy crisis of the 1970s described in Chapter 8, the climate change issue has provoked considerable debate about both energy futures and energy policy.

There is already considerable agreement about the likely framework for a Climate Convention. The precedents set by the 1979 LRTAP and 1985 Vienna Conventions are likely to be built upon, with a 'framework' Convention, specifying broad principles and procedures, being established initially. These general obligations would be designed to be acceptable to the greatest number of countries. Specific obligations would be achieved through protocols (cf. the Helsinki, Sofia and Montreal Protocols) which would be the subject of further negotiations. Not every signatory to the Convention may find every Protocol acceptable, as has proved to be the case with the UK and the US failing to sign the Helsinki Protocol on sulphur dioxide (SO_2) emissions. The possible subjects of protocols include reduction targets for greenhouse gas emissions, technology transfer and financial assistance arrangements for developing countries.

The challenge of negotiating such a Climate Convention is immense. Among the most difficult problem will be securing the support of developing countries which see climate change as the consequence of the past activities of the developed world. However, the likelihood that China and India will sign the Montreal Protocol on CFC emissions in exchange for increased financial support is a promising sign. A growing difficulty is US reluctance to support reductions in greenhouse gas emissions.

The major importance of the climate change issue in both the UK and the FR Germany is discussed below. There are some signs that the European climate change debate might follow a path similar to that for acid rain, with Britain calling for realism in the run-up to a Climate Convention and Germany for precautionary policies, speedy action and faster promotion of technological solutions. Policy responses to climate change are likely to have major implications for the energy supply industries. These are discussed in relation to Britain and the FR Germany in the following sections of this chapter.

Energy/environmental issues in Europe

New acid rain developments

While climate change is now the dominant environmental issue in Europe, the scientific, political and diplomatic processes with respect to the acid emissions have not ceased. In most countries, acid-rain damage to buildings and surface waters is now accepted as a scientific fact. However, the question of impacts on trees remains controversial.[3]

If new international agreements are to be struck, then new concepts being developed within the framework of the UNECE LRTAP Convention are likely to be of some importance. In particular, signatories to the Sofia Protocol, including the UK and the FR Germany, have accepted the concept of critical pollution loads as the basis for future negotiations.

A critical pollution load may be defined as a 'quantitative estimate of an exposure to one or more pollutants below which significant harmful effects on specified sensitive elements of the environment do not occur according to present knowledge'.[4] The concept has analogies with the dose-response relationship used in assessing the effect of pollution on human health. At the national level, the critical load concept offers the potential to apply abatement technologies when and where they are most effective in minimizing impacts on sensitive ecosystems. It also requires the active participation of scientists in the regulatory process.

The critical load concept has appeared attractive to the British government because it is consistent with the stated policy of basing policy on science and it promises a cost-effective use of resources. There has been less discussion of the concept in the FR Germany where policy has been driven more by technology and where substantial reductions in emissions have, in any case, been made.

There is also a view that the very complexity of the critical load as a control instrument makes it unlikely that it will be enshrined in international legislation 'except perhaps as a stimuli ... to intensify collective efforts and provide scientific justification for the expenditure of large sums of money'.[5]

While detailed work on estimating critical loads for specific ecosystems is now under way, initial estimates tend to suggest that substantial reductions in European emissions would be necessary to secure given environmental objectives. This was the conclusion of the UK's Acid Waters Review Group.[6] This could have significant implications for future UK abatement policy in particular given the relatively low level of planned emissions reductions.

Related to the concept of the critical load is that of the 'target load'. This is not the property of a particular ecosystem but it embodies other factors such as economic or political acceptability. Although as a matter of political realism it is likely that the target loads used to drive any abatement policy will be more permissive than the critical loads, there is no reason in principle why they should not be more restrictive if a margin of error were sought to protect against the possibility of environmental damage.

Control of stationary sources

While most European countries have now entered into some form of agreement on acid emissions, either through the EC or UNECE, the amount of active emissions abatement carried out varies widely. FGD has been the primary technique used for abating SO_2 emissions since the 1970s. Table 13.1 shows the amount of FGD plant installed in different countries. Europe accounts for about 35 per cent of the FGD currently in place round the world, with the FR Germany making up 95 per cent of the EC total. Over the coming decade, the distribution of FGD installations will become much more widespread.

However, FGD is becoming less of a key component in SO_2 emissions abatement programmes. In both Europe and the US, the use of low sulphur coal is likely to play a much larger role as is, for new installations, the use of natural gas burned in efficient combined cycle stations.

NO_x controls are also becoming more widespread. However, most countries are relying on combustion modification to secure modest emission reductions. The use of expensive selective catalytic reduction (SCR) technology which removes 80–90 per cent of the NO_x in flue gases is restricted to Japan and the FR Germany.

Vehicles emissions control

With power stations controls now in place, the transport sector will attract an increasing amount of attention as reductions in the environmental impact of

Table 13.1   Coal-fired power plant capacity fitted with FGD (MW$_{the}$)

	In place	Firmly placed
FR Germany	119830	33200
UK	–	23900
Austria	5460	1050
Canada	–	13800
Denmark	–	4500
Finland	–	460
Italy	–	25530
Japan	31600	3750
Netherlands	7500	3600
Spain	–	1100
Sweden	3200	–
Taiwan	–	6200
US	212900	111900
European Community	127330	91830
Europe	135990	93340
World	383640	228990

Source: IEA Coal Research 1989.

fossil fuel use are sought. Transport policy itself rather than the technical characteristics of vehicle engines may become the primary focus, as is already occurring in the FR Germany. However, most of the recent European debate has focused on exhaust emissions standards.

The EC has increasingly become the forum within which Europe-wide emissions standards are set. The story of vehicle emission regulation in the EC, which ran parallel to that of the regulation of stationary sources and was no less convoluted, is told briefly here.

Until 1983, the EC had simply incorporated relatively permissive UNECE standards directly into Community law. However, in 1984, the Commission was persuaded by Germany to propose the equivalent of US standards for new vehicle types. As was the case with the LCP Directive, some of the most vigorous opposition came from the UK. A major concern for the UK government was the threat to the nationally owned British Leyland car group which had opted for lean burn technology. Ford Europe, with its dominant position in the smaller, lower-value end of the car market, was also opposed to the adoption of catalytic convertors given the potential impact on manufacturing costs and the threat of Japanese competition.

Between 1984 and 1988 a long European battle between governments, the car industry and environmentalists took place. A temporary compromise was reached in 1985 ('the Luxembourg agreement') setting standards, which would allow lean-burn engines only for small cars with engines sized below 1.4 litres. Medium-sized cars (1.4–2.0 litres) would require the use of cheaper oxidation catalysts in combination with lean-burn engines. However, the

Luxembourg agreement did not become law until December 1987 when two objecting Member States, Denmark and Greece, were outvoted under qualified majority voting permitted by the Single European Act.

In June 1988, a further deal was reached whereby exhaust standards for small cars would be tightened further, requiring the use of an oxidation catalyst. Individual Member States would, however, have been allowed to offer fiscal incentives for cleaner cars. The French subsequently withdrew their consent from this agreement after pressure from Peugeot which objected to the fiscal incentives. A November 1988 agreement which omitted the fiscal incentives fell apart after rejection by the European Parliament and a Dutch promise to go it alone.

Finally, in June 1989, the Council agreed a Commission recommendation for emission standards for small cars which would require three-way catalytic convertors. The Member States with substantial small-car manufacturing capacity – the UK, Spain, France and Italy – had by then bowed to the inevitable. With British Leyland privatized, the UK government gave up its defence of the lean burn engine after the extraction of a promise that the fuel efficiency of cars would be the subject of future EC negotiations.

This agreement left a huge anomaly in the EC's regulatory system because small cars were by then subject to tighter standards than medium-sized and large ones. A 'politically binding' promise was made that the tighter standards would soon be applied to all cars. This is being given more formal shape by the European Commission. After 1992, all new cars sold in the EC will be equipped with best practice emissions abatement technology.

Energy policy in Europe

Energy is a policy domain which has been guarded jealously by EC Member States. However, there have been a number of developments during the latter part of the 1980s which are relevant to environmental policy. The first was the formulation, in 1986, of a set of strategic objectives for the year 1995 by the Council of Energy Ministers. The Council called for the Commission to bring forward proposals which would 'increas(e) the convergence and the cohesion of the Member States' energy policies'. The second development was the objective of completing the internal energy market as part of the 1992 initiative. There are, however, areas of tension between energy and environmental policy. In addition, there are possible tensions between the '1992' initiative and the strategic objectives defined in 1986.[7]

The 1995 objectives included:

1   the achievement of greater energy efficiency, specifically a 20 per cent improvement by 1995;
2   limiting oil imports and, specifically, keeping these to less than one-third of energy consumption by 1995;
3   increasing the share of solid fuels in energy consumption; and
4   reducing the share of hydrocarbons in electricity production to less than 15 per cent by 1995.

The Council of Ministers avoided direct reference to the issues of nuclear power because of fundamental differences in the views of Member States. The 1986 Council Resolution was made immediately after the Chernobyl accident.

Progress towards the achievement of these goals is patchy. In particular, low energy prices have reduced incentives for energy efficiency and it is likely that the share of solid fuels in energy consumption will fall rather than rise.

There are also tensions with environmental policy objectives. The LCP Directive has significantly reduced the attraction of solid fuels while a combination of low prices and environmental controls has significantly enhanced the market position of natural gas, one of the hydrocarbon fuels, particularly for power generation. Future policies to cut CO_2 emissions may well encourage gas consumption at the expense of coal while providing additional arguments for those who favour the development of nuclear power.

The 1992 objectives articulated by the Commission[8] include:

1   removing obstacles which already breach Community law, including restricted movement of goods, State monopolies and State aids. Given Member States' past jealousies over energy policy such obstacles abound, Germany and the UK being no exception;
2   transparency and uniformity in energy tariffs and prices;
3   encouraging the development of the energy infrastructure, particularly gas and electricity grids; and
4   reconciling the need for uniform environmental standards with the right of individual Member States to adopt stringent measures of their own.

The biggest obstacle to the Single Energy Market in Britain and Germany has probably been aid to the coal industry. However, government support for the British nuclear industry has also been defined as a State Aid which must be removed by 1998. In addition, the German electric utilities' practice of not allowing third parties common carriage rights on their transmission networks is also a problem. Electricité de France has already complained about its inability to supply large German industrial consumers just over the border.

Given the economic liberalism of the Thatcher administrations, UK domestic energy policy is fundamentally in tune with the 1992 objectives. The Federal Government in Germany has ambitions in the same direction but the electric utilities jealously guard their territorial rights. Co-operative business practices, both within the electricity supply industry and with the coal industry, are common.

Laissez-faire energy policies, encouraged by the 1992 objectives are, at the moment, promoting the greater use of natural gas which will reduce both acid emissions and those of CO_2. However, this convergence of policy objectives may only be a temporary consequence of present low gas prices. In the longer term, a more distant relationship between governments and utilities may reduce the levers of power available to regulators and make environmental policies, particularly if these involve the promotion of end-use energy efficiency rather than supply-side technological fixes, more difficult to implement.

Global warming

The EC has begun to take a major interest in the question of global warming within the context of its Fourth Action Programme. Because responses to climate change would require significant changes to patterns of energy supply and demand, the Environment Directorate of the Commission has been operating in concert with the Energy Directorate. In November 1988, the Commission made a formal Communication to the Council of Ministers in which it called for a virtual ban on CFC production by the year 2000, Community support for a global Climate Convention and the setting of a £4m research programme to assess different options for mitigating and adapting to climate change.[9]

In 1989, the Energy Directorate issued a set of energy and emission projections which foresaw EC CO_2 emissions continuing to rise without policy action. The Directorate believed that 'energy efficiency improvements seem to be the most promising single policy tool'.

In March 1990, the Commission proposed that the EC should undertake to stabilize its CO_2 emissions at their 1990 level by the year 2000 with, however, derogations for Greece, Spain, Ireland and Portugal in view of their lower degree of industrialization. This proposal was criticized by the UK. The Prime Minister subsequently suggested a target for the UK of a stabilization by the year 2005. The FR Germany believes that a reduction of CO_2 emissions of 20–30 per cent is possible by that time, indicating that negotiations on CO_2 emissions within the Community may be determined by the same fundamental differences of interest made evident during discussions on the LCP Directive.

Developments in Britain

Acid rain

The fact that surface waters have been acidified by air pollution is widely accepted in the UK, though there is less certainty about damage to trees and buildings. While at the scientific and official level there is an understanding that acid rain is damaging the natural environment in the UK,[10] widespread public concern is still lacking. In a 1989 DoE opinion poll,[11] acid rain ranked 11th on a list of environmental problems well behind sewage contamination, radioactive waste disposal, chemicals in rivers, rainforest loss, ozone depletion and climate change. As discussed in Chapter 4, the environmental assets damaged by acid rain have little economic value, are not highly valued by the British public and are located in remoter areas, well away from centres of population.

The greening of the government

Since the agreement of the LCP Directive in June 1988, the UK Government has made an effort to shake off the 'dirty man of Europe' image and adopt a more positive stance on environmental issues. The turning point came in a speech made to the Royal Society by Mrs Thatcher in September 1988 in which she raised the possibility that man had 'unwittingly begun a massive experiment with the system of this planet itself.' This, the first serious reference to the environment made by the Prime Minister in nine years of office, has been likened to the conversion of Saul on the road to Damascus.

The government's interest in the environment was linked to the growing salience of global environmental issues such as climate change and the depletion of the ozone layer rather than to direct European pressures. Only a day after Mrs Thatcher's speech, the Foreign Secretary called on the UN General Assembly to have a debate about the climate change issue. After years of neglect of environmental issues, the Royal Society speech raised high expectations about future policy initiatives.

However, there was also cynicism among the environmental movement about the motivations for the speech. At the 1988 Conservative Party Conference, the Prime Minister advocated the 'safe, sensible and balanced use of nuclear power' as a solution to global warming, raising suspicions that one of the motivations underlying the Royal Society speech had been the promotion of nuclear power.[12] In November 1989, the government rejected the recommendations of the House of Commons Energy Committee that immediate priority should be given to improving energy efficiency through 'a mixture of regulation, penalties and incentives'.[13]

The government's greening has helped to raise public consciousness about environmental issues considerably. As noted in Chapter 4, the environment jumped from being the eighth most important national problem in a 1986 Department of the Environment opinion poll (mentioned by 8 per cent of respondents) to second place in 1989 (mentioned by 30 per cent).[14] In June 1989, the Green Party won an extraordinary 15 per cent of the vote in elections to the European Parliament. In response, the Prime Minister replaced the Environment Secretary Nicholas Ridley, as undiplomatic about the environmental movement as he was later to be about Germans, with the politically more acceptable Chris Patten.

Mr Patten has put the concepts of sustainable development and taxing industry and consumers to reflect the external costs of environmental pollution firmly on the political map in the UK. He gave great prominence to a report on sustainable development, later turned into a popular paperback,[15] which had been commissioned by the DoE.

However, the government's performance on more mundane environmental problems has detracted from its rhetoric on global issues. In 1990, the European Commission decided to prosecute the UK over the quality of water at a number of bathing beaches, ending the UK's claim that, while it might argue hard against the adoption of certain environmental measures, it would always comply with its legal obligations.

In September 1990, the government produced a White Paper on future environmental strategy.[16] Although it had been widely anticipated that the paper would mark a significant re-appraisal of government policy, both environmentalists and independent commentators expressed disappointment at the way in which options for action were discussed only in general terms. In particular, no strategy for achieving the Prime Minister's objective of freezing CO_2 emissions in the UK by the year 2005 was annunciated. There appears to be little likelihood that environmental taxes, apart from in the transport sector, will be introduced in the near-term. There was a widespread belief that the Treasury had vetoed any moves towards environmental taxes because of the potential effects on inflation and the privatisation of the electricity industry. Nevertheless, the first chapter of the White Paper signals a new willingness on the part of the government to accept a greater role for regulatory measures to limit environmental impacts. This may have implications for policy later in the 1990s.

Institutional changes

While the UK has developed a highly visible superstructure for its environmental policy, the present deep malaise in the new inspection body, HM Inspectorate of Pollution (HMIP), symbolizes years of neglect of the regulatory infrastructure which has weakened the country's capacity to respond to new environmental challenges.

HMIP was created in 1987 as a component part of the DoE, bringing together the Industrial Air Pollution Inspectorate (IAPI), the Radiochemical Inspectorate, the Hazardous Waste Inspectorate and a new water inspectorate. The amalgamation was on the recommendation of a 1986 Efficiency Scrutiny Report on industrial inspection.[17] The Scrutiny team focused narrowly on organizational issues and did not mention the broader context of increasing responsibilities for pollution control in the light of EC obligations. A recommendation that 'a substantial part of IAPI's resources should be diverted into other sectors of pollution control' was accepted in principle but did not, in practice, take place.

An important function of HMIP, suggested by the Royal Commission on Environmental Pollution as early as 1976, is to be the development of a system of Integrated Pollution Control (IPC) which will take account of emissions to all media, including air, water and solid waste disposal. HMIP has carried out two case studies of IPC as it might be applied to individual sites and has concluded that it could move to a system where a single inspector was responsible for all of HMIP's interests at a given site.[18] This would involve considerable re-training and internal re-organization as there are still major cultural divisions between the different classes of inspector within HMIP.

The first three years of HMIP's experience have been dominated by the problems of understaffing, personnel difficulties, recruitment and work overload. By the end of 1989, three of HMIP's most senior inspectors had resigned and its Director had committed suicide.[19] When HMIP was formed it was 66 short of its complement of 214 posts. In spite of a number of recruitment drives, there were still more than 20 vacancies in early 1990. Low salaries compared to industry are a major part of the problem. A new post of Assistant Inspector with less stringent qualifying requirements is being created.

There have been major difficulties associated with integrating the component inspectorates which formed HMIP. The air inspectorate had developed a generally cooperative relationship with regulated firms while the radiochemical inspectors had a much more arms-length approach to the nuclear industry. The integration of inspectors into Civil Service pay scales resulted in a much resented loss of status. Regional re-organization within HMIP also helped to alienate the air pollution inspectors. In 1990, the government decided to make HMIP a candidate for agency status, removing it from the DoE.

Air inspectors have recently had the responsibility of preparing new 'best available techniques not entailing excessive costs' (BATNEEC) notes for processes which have been scheduled as a result of the EC's 'framework' Directive on emissions from industrial plants. The new 'large combustion plants' class of installations, which replaces the 'electricity works' class, will comprise many more sites than those previously scheduled. Not surprisingly, site inspection rates have been declining, bringing into question once again the effectiveness of industrial inspection.

The Environmental Protection Act

Legislation revising the UK air-pollution control system to allow the implementation of European obligations was promised by the government in 1986 but failed to find Parliamentary time. Even following the Prime Minister's 1988 speech, a 'Green Bill' which was expected in the 1989–90 Parliamentary Session failed to appear.

The Environmental Protection Bill was finally introduced into Parliament in December 1989. It proposes a thorough overhaul of British environmental legislation and, in many areas, introduces clear consistency with European obligations for the first time. The Bill gives a legal definition of the IPC concept, the application of which is to be the responsibility of HMIP. The Bill was enacted in November 1990, several months behind schedule.

There are a number of innovations on the air-pollution side which will help the UK to implement the LCP Directive and which may provide the basis for some action to abate greenhouse gas emissions. The DoE is empowered to specify directly legally binding limits on the concentrations or the total quantities of prescribed substances such as SO_2 or NO_x emitted in any time period. Such regulations may discriminate according to type of process, owner or locality. This provision will allow the emission limits for new plant covered by the LCP Directive to be implemented.

The DoE may establish air-quality standards or objectives which must be taken into account by HMIP and the local authorities when granting authorizations for individual plants.

The DoE may also 'make plans for establishing limits for the total amount in any period, of any substance which may be released from prescribed processes into the environment in, or in any area within, the UK'. These plans may allocate quotas to operators of prescribed processes and may specify the progressive reduction of allowed emission levels. This authority will allow the aggregate national emission reduction targets for existing LCP to be met.

The former system of annual registration of scheduled processes is to be replaced by a system of prior authorization as required by the EC 'framework'

Directive. Regulations may specify whether a process is to be controlled by HMIP or by local authorities. In granting an authorization, the control authority must take into account: EC or other international law; air quality objectives; and the requirements of any emissions reduction plan specified by the DoE.

The fundamental principle to be applied by HMIP or local authorities in granting authorizations for scheduled industrial plant will be that of 'best available techniques not entailing excessive cost' (BATNEEC) as opposed to the traditional BPM. British BATNEEC is subtly different from EC BATNEEC in that 'techniques' rather than 'technology' is specified. It has been argued that this is a deliberate weakening of the EC's 'framework' Directive. The obligation to use BATNEEC falls not only on the control authority, but also directly on the operator of a scheduled process.

The powers in the Environmental Protection Act provide for a flexible and comprehensive system of environmental controls within which regulations of any degree of stringency, covering new or existing plant, might be applied. The general regulatory philosophy sits perhaps uneasily with the government's broader political rhetoric about market-based environmental incentives, for which there is no explicit provision.

Electricity privatization

The privatization of the electricity supply industry in Britain represents a unique experiment which is being watched with interest in many other countries. A key premise underlying the privatization is that, whereas electricity transmission and distribution are natural monopolies, it is possible to create a competitive market for electricity generation. Since generation is solely responsible for the power industry's atmospheric emissions, this has implications for environmental regulation.

The privatization process has been extremely complex and arrangements have changed substantially during the two years of preparation. The two principal features relevant to environmental controls are: 1) industry structure; and 2) the arrangements for the new electricity market. The arrangements for England and Wales only are described.

The government originally proposed splitting the Central Electricity Generating Board (CEGB) into three parts in its February 1988 White Paper.[20] These were a transmission company (now known as National Grid); a generating company with 70 per cent of the capacity including all the nuclear plant (now known as National Power); and a smaller company with the remainder of the assets (now known as PowerGen). The obligation to supply electricity was to be taken away from the CEGB's successors and given to the regional distribution companies, known now as 'regional electricity companies' (RECs).

This revolutionary structure could not be further from the traditional model of vertically integrated, territorially defined utilities, subject to financial regulation and with an obligation to supply, as found in the FR Germany.

The twin desires of the government were to introduce competition and to promote the development of nuclear power. The government was advised early on that these objectives were unlikely to be compatible.[21] To introduce competition, it was necessary to have more than one generating company. However, to absorb in a single company the large liabilities associated with decommissioning nuclear power stations, the nuclear assets had to be diluted with a large amount of fossil capacity. This line of argument led inexorably to the creation of National Power.

In any event, it became evident that the CEGB's existing nuclear assets were not marketable to investors and that new nuclear capacity was likely to be uneconomic. In 1989, the CEGB's nuclear assets were withdrawn from the electricity sale and a new, nationally owned company, Nuclear Electric, was proposed. At the end of March 1990, National Power and PowerGen were vested, owning 60 per cent and 40 per cent respectively of the former CEGB's fossil assets.

The creation of a market in bulk power supplies was complicated by the simultaneous desires to promote competition, establish contracts between electricity generators and distributors and continue the traditional system of least-cost dispatching of generating plant. Because the obligation to supply was to rest with the distribution companies, these constraints led to the construction of a highly complicated double power pool system where the distributors pooled their electricity contracts and the generators pooled their power plants. This proposed system was abandoned as unworkable in mid-1989, but only after a complete computer-based pooling and settlements system had been developed.

In the end, a simpler spot market for electricity has been established. There are no formal contracts between electricity generators and RECs and the obligation to supply electricity has been abandoned altogether. The incentive to build power stations is supposed to come from the inclusion, in the price at which electricity is traded in the spot market, of an element related to the risk of power cuts.

Environmental implications of privatization

The privatization has had a rapid effect on the investment policies of the electricity supply industry as the generating companies, exposed to full financial risks, have adopted the 'short-termism' typical of most sectors of British industry. Capital-intensive nuclear power has been abandoned, as has the construction of large new coal-fired power stations. With the price of gas low, cheap combined cycle gas turbine plants will account for most of the new generating capacity installed during the 1990s.

At the same time, free to pursue their own fuel procurement policies, the generating companies are likely to increase their consumption of imported coal which is relatively low in sulphur. The requirements of the LCP Directive, as implemented through the Environmental Protection Act, have enhanced the attractiveness of this option.

Initially, the new technological trajectory of the privatized industry will tend to yield environmental benefits due to the greater use of low sulphur coal and natural gas. However, this is partly due to the short time horizons presently being used to appraise investments. Towards the end of the 1990s, if the price of imported energy rises, technical fixes such as FGD may appear, in retrospect, to have been more attractive.

In the mean time, the financial disciplines of competing in a spot market for power are likely to make the CEGB's successors highly resistant to any attempt to tighten up requirements for emission reductions from existing plant. It is also possible that clean plant fitted with FGD could have its output reduced as a result of higher costs and competition from uncontrolled stations.

In the longer term, if environmental benefits are sought from the improved efficiency of electricity use, there are major obstacles within the new system of price regulation applied to the RECs. The price formulae used by the Office of Electricity Regulation (OFFER) allow significantly increased profits for those who are able to expand power sales. While this situation persists, the RECs will always be reluctant partners in any policy initiative aimed at reducing electricity consumption.

Developments in Germany

Acid rain

Forest dieback has now come to be seen as a much more complex phenomenon than had at first been assumed. It is now believed to be attributable to a number of stresses, not all of which are well understood. The idea that air pollution significantly contributed to this stress is, nevertheless, accepted as self-evident. The German government has done what it can for the forest and has ensured that the protection of forests from air pollution is now an EC concern.

The issue has therefore been allowed to slide from the political limelight. However, the Cabinet agreed in 1988 to continue the Action Programme to Save the Forest as a tool for forestry research and management. The Federal Environment Office (UBA) has continued to promote further efforts to restrict acid emissions from the transport sector which now presents the greatest challenge in the field of air pollution. There is a continuing emphasis on reducing NO_x emissions, including those from commercial vehicles and aircraft.

Tax incentives for clean new cars are still in place and those for small cars were increased in 1990. However, there has been disappointment that car owners have not retrofitted old cars with catalytic converters, a system of financial incentives to encourage the retrofitting of old cars therefore began in January 1990. Bonn has meanwhile been pressuring Brussels to tighten emission standards for heavy vehicles, justifying this with reference to the Sofia Protocol on NO_x emissions.

Environmental policy

Until unification issues swept across Germany in late 1989, there were no signs of energy and environmental issues becoming less salient in the FR

Germany. These policy areas have, however, retained a high political profile because of the problems now faced in the East as well as the continuing market competition between the various energy sources, especially coal, nuclear power and, more recently, natural gas.

The political parties continue to compete for green laurels. The two main political groupings, the CDU/CSU and the SPD, have each produced documents intended to place their environmental policies on a firmer philosophical basis. The CDU/CSU appealed to Christian ideology, entitling their programme 'Man's responsibility to the Creation'. The SPD's programme refers to the need for an 'ecological restructuring of the economy'.

It is the SPD's policies which are now most informed by green ideas. The former Chancellor, Helmut Schmidt, has stressed the global risks associated with present energy use patterns and had pleaded for massive investments in an energy sector free of both nuclear power and fossil fuels. For the short term, he has recommended a continuing reliance on a diversity of energy sources coupled with energy saving and higher energy prices. However, for the longer term, he has advocated an essentially green energy scenario.[22]

A debate about environmental taxes exploded in Germany in the summer of 1989, although there had, in the past, been a fairly weak response to proposals of this kind. Both the CDU/CSU and the SPD have proposed such policy instruments. However there are major differences in their approaches. The SPD, for example, reflecting its constituency in North Rhine-Westphalia, proposes exempting coal from an energy tax. To many, this would appear to defeat the purpose of such a tax. Both the energy industries and consumers in East Germany have an interest in this debate.

German industry views environmental protection and 'green' product quality as an essential ingredient of marketing strategies and an important competitive factor, especially after 1992. However, industry also continues to argue that competitors abroad should bear similar costs. Chancellor Kohl has responded to concern about the impact of CO_2 emission reductions by urging uniform measures for Europe 'with the aim of spreading the load of possible competitive disadvantages.'[23] Industry also claims that forward planning and investment has been made difficult because of the increasing rate at which new regulations are coming on to the statute book.

By 1990, the unification of Germany and the growing clamour for Western economic assistance from other East European countries had posed new economic opportunities and environmental challenges. The German government faced the huge task of persuading its taxpayers to subsidize investment in the East for both environmental and political reasons. Environmental imperatives now provide a major justification for West German interest in reducing the use of high sulphur lignite in the East.

New legislation

There has been a continuing emphasis on developing new legal concepts, more effective instruments and better ways of operationalizing the precautionary approach to environmental protection, encompassing the use of both state-of-the-art technology and non-technical measures.[24] Environmental protection has become one of the more important rationales for investing in and formulating policies for the East. Once again, expanding markets and the environmental cause seem to be going hand in hand. There is a wide consensus that it is the role of government to make certain that technology and products continue to evolve.

The pace of legislative change has clearly not slackened since the Federal Air Quality Protection Law was amended in 1985 and the Technical Instructions on Air – TA (Luft) – were updated in 1986. By 1990, five legislative initiatives were under way, at least three with implications for the energy industries:

1   a new Federal Act on Environmental Compatibility Assessment which implements the EC's Environmental Impact Assessment Directive;
2   the further tightening of the Federal Air Quality Protection Act and the revision of the Small Combustion Plant Regulation;
3   a new Chemicals Act;
4   a Federal Waste Water Charges Act, which ensures that polluters pay more for the liquid wastes which they discharge; and
5   amendments to the Nature Protection Act, which has already brought the Federal Environment Ministry (BMU) into conflict with the Agriculture Ministry (BMELF) and wider agricultural interests.

There is still an official ambition to assist small firms and consumers in adopting more environment friendly practices. The government continues to support, with tax concessions and low-interest loans, firms and consumers who buy environmentally friendly products and invest in approved technologies.

Energy sector

The politicization of German energy policy was undiminished by 1990. Major issues on the agenda still include the future roles of nuclear power and coal as well as the structure and regulation of the electric utilities.

Green opposition to nuclear power has now extended to the other parties. As early as 1984, the SPD decided to phase out nuclear power gradually and this has now been established as a fundamental policy objective. An anti-nuclear caucus has operated within the traditionally pro-nuclear CDU since 1988. The CDU/CSU/FPD government has abandoned plans to reprocess nuclear waste at Wackersdorf in Bavaria and is, instead, sending the waste to France and the UK. However, in spite of widespread public and political opposition to nuclear power, the German nuclear industry continues to make its case, placing much reliance on the potential contribution which nuclear might make to reducing CO_2 emissions and cleaning up the East. A Secretary of State at BMU has endorsed the view that nuclear power will have a role to play in addressing the climate change problem.[25]

Strong European pressure is being exerted on Germany not to review the Jahrhundertvertrag ('century contract') which still underpins the use of hard coal in the German utility sector. The contract is due to expire in 1994. This, and the Kohlepfennig ('coal penny'), which accounts for 8.5 per cent of German electricity prices, are seen to be obstacles to competitive energy markets. The Federal Economics Ministry (BMWi) is also keen to see the Kohlepfennig go, as are some of the Länder. The coal industry and the SPD remain opposed to such a move on energy security and social grounds.

If the Kohlepfennig were to disappear, markets for hard coal in West Germany could decline substantially. Whether new markets might open up in the East is one of the issues which is being addressed with unification.

Should the use of hard coal for electricity generation decline substantially after 1994, there might well be a question mark over the worth of part of the massive 1983–8 FGD programme. If FGD plants operate for less than ten years and investment costs are written off early, the implied cost of SO_2 abatement will have been very high indeed. On the other hand, the very existence of coal-fired stations fitted with FGD may help to protect coal's market against the encroachment of natural gas.

In 1994, the concessionary contracts defining the electric utilities territorial rights are up for renewal. This opens up the possibility of substantial changes in the role of the utilities and their relationships with each other. A more competitive structure with provision for common carriage would be supported by BMWi and the European Commission. The utilities themselves are bitterly opposed to such changes and the Association of German Electricity Producers (VDEW) has strongly asserted that electricity supply is a natural monopoly and that the questions of the obligation to supply power and territorial concessions are intimately interlinked. The Green Party's proposals to amend the 1935 Energy Sector Act are the subject of serious debate even within the SPD. US concepts, such as least-cost planning to promote the efficient use of electricity and regulatory intervention in favour of renewable energy and independent power generation, are being widely discussed.

Energy and environment research remains a priority for the Federal Research and Technology Ministry. For 1990, DM 5bn was allocated. Among the aims is now the promotion of strategies to reduce greenhouse gas emissions. There are also DM 245m set aside for clean coal technology and DM 294m for renewables and efficient energy use. This element is expected to rise in the early 1990s. However, expenditure on nuclear-related research remains high at over DM 600m. BMFT clearly remains primarily a nuclear research ministry.[26]

Climate change

Public discussion of the climate change issue began earlier than in Britain, *Der Spiegel* publishing a major article as early as 1985. The nuclear industry started using climate change as a rationale for the development of nuclear power at about the same time.

One of the most important developments has been the recall of the Enquete Commission on Energy and the Environment in order to undertake a year-long Inquiry into the 'Protection of the Earth's Atmosphere'. The Commission, which consists of both Parliamentarians and scientists, produced its interim report in early 1989, with the final report scheduled for late-1990. The scope of the Inquiry is enormous, covering energy/emission projections, impacts of climate change and international issues relating to the negotiation of a Climate Change convention. Work sponsored by the Commission as part of the Inquiry has kept large sections of the German research community active.

BMU supports the stabilization of CO_2 emission in highly industrialized countries by the year 2000 and has urged a 25 per cent reduction by 2000 for the Federal Republic. Chancellor Kohl has backed this call. Already Germany has urged CO_2 emission reductions of that order on its EC partners. As was the case for acid emissions, it is clear that the FR Germany will be taking a leading role in pressing for precautionary measures in anticipation of major climatic changes. This will have major implications for both EC Member States and other European countries.

Notes

1. World Commission on Environment and Development (1987), *Our Common Future*, Oxford University Press.
2. A minority of scientific opinion dissents from this consensus view. Scepticism has, in particular been expressed by the US-based George C Marshall Institute.
3. Skeffington *et al.*, (November 1988), *Nature*, 1 (339), No. 6194; Schulz, 1989, Wie kann dem Wald geholfen werden?, *Mitteilungen der Deutsche Forschunggemeinschaft*, 3/1989; and P. Schutt and E.B. Cowling (1985), Waldsterben: a general Decline of Forests in Central Europe, *Plant Disease*, **69**, (7).
4. Sofia Protocol (1988), Article 1, para. 7.
5. N. Haigh (1989), New tools for European air pollution control, *International Journal of Environmental Affairs*, Vol. 1, No. 1, pp. 26–37.
6. UK Acid Waters Review Group (December 1988), *Acidity in UK Fresh Waters*, 2nd report, HMSO, London.
7. F. McGowan (December 1989), The Single Energy Market and energy policy, *Energy policy*, Vol. 17, No. 6, Butterworth, Guildford, pp. 547–53.
8. Commission of the European Communities (1988), *Commission Working Document: the Internal Energy Market*, COM(88) 238 final, Brussels.
9. Commission of the European Communities, *Communication to the Council: The Greenhouse Effect and the Community*, COM(88) 656 final, Brussels, 16 November 1988.
10. UK Acid Waters Review Group (December 1988), *Acidity in UK Fresh Waters*, 2nd report, HMSO, London.
11. Department of the Environment (1990), Public attitudes to the environment, *Digest of environmental protection and water statistics*, No. 12, Part 10, HMSO, London.
12. Environmental Data Services Ltd (October 1988), *ENDS report 165*, London, p. 21.
13. House of Commons Energy Committee (July 1989), *Energy Policy Implications of the Greenhouse Effect*, Sixth report, Session 1988–9, HMSO, London.
14. Department of the Environment (1990), Public attitudes to the environment, *Digest of Environmental Protection and Water Statistics*, No. 12, Part 10, HMSO, London.
15. D. Pearce *et al.* (1989), *Blueprint for a Green Economy*, Earthscan, London.
16. *This Common Inheritance: Britain's Environmental Strategy*, (September 1990), HMSO, London.
17. Efficiency Scrutiny Report (1986), *Inspecting Industry: Pollution and Safety*, HMSO, London.

18. Department of the Environment (1989), *HM Inspectorate of Pollution: First Annual Report 1987–88*, HMSO, London.
19. Environmental Data Services Ltd (November and December 1989), *ENDS Reports 178 and 179*, London, pp. 11–15 and 3.
20. February 1988, *Privatising Electricity*, Cm 322, HMSO, London.
21. House of Commons Energy Committee (1988), *The Structure, Regulation and Economic Consequences of Electricity Supply in the Private Sector*, 3rd report, Session 1987–8, HC 307, HMSO, London.
22. H. Schmidt (19 February 1988), Sieben Prinzipien vernunftiger Energiepolitik, *Die Zeit*.
23. Financial Times Busines Information (20 July 1989), Environment top priority in German energy policy, *Power in Europe*.
24. Klopfer and Kroger (1990), Zur Konkretisierung der immisions-schutzrechtlichen Vorsorgepflicht, *Natur und Recht*, Heft 1.
25. Clemens Strotmann, Secretary of State at the Federal Environment Ministry, Speech to the Society for Reactor Safety, reported in *Umwelt*, 2/1990.
26. Financial Times Business Information (9 March 1990), *European Energy Report*, No. 309.

14 Acid rain policy in Britain and Germany: explanation and evaluation

Introduction

The rationality and the effectiveness of policy are necessarily relative and depend on the context within which they are judged. Any evaluation of environmental policy which aspires to a measure of objectivity must, therefore, make its criteria clear. In practice, depending on ideology, priorities and circumstance, governments may take different views about which of a wide range of criteria are most appropriate.

In the energy/environment policy field, policy-makers may legitimately take into account impacts on different energy industries, economic growth and technological capability as well as environmental damage itself. Environmental measures may be judged as much by their contribution to these broader policy goals as by their effectiveness in furthering environmental improvement.

In the acid-rain debate, the perspectives of Britain, Germany and the Scandinavian countries have differed, as have the perspectives of the bureaucracies charged with formulating and implementing energy or environmental policy at both the national and EC levels.

Nevertheless, leaving aside the incomplete nature of current emission abatement programmes and the considerable residual uncertainty about certain types of damage attributed to acid rain, the conclusion reached in this book is that German and British policies during the 1980s were rational in the sense that the governments in each country succeeded in furthering their own interests viewed from a wider political and economic perspective.

Consequently, given a very diverse set of political objectives and a lack of consensus about the scientific base of knowledge on which decisions were based, it may be that EC measures on power station and vehicle emissions negotiated during the 1980s were the best achievable at political costs which were acceptable to the individual negotiating partners. Taking a broader European perspective, a more expeditious and less conflict-ridden resolution of national differences would however be desirable and should be possible in the future.

This conclusion may not satisfy those for whom environmental priorities are pre-eminent. It does however illustrate that those who wish to further environmental goals may need to consider the compatibility of their proposals with wider policy goals. In particular, the possibility of environmental regulation being pursued by governments in order to achieve non-environmental objectives may need to be recognized and, where necessary, exploited.

Acknowledging these broad conclusions, this concluding chapter therefore:

1 isolates the ingredients of environmental policy-making in Britain and Germany by considering the respective roles of science, technology and perceived economic impacts;
2 identifies differences in the decision-making contexts in Britain and Germany which offer more general explanations for the divergences in policy; and
3 evaluates the success of German and British acid-rain policies against environmental, economic, technological and political criteria.

The ingredients of environmental policy

Science and the perception of environmental impacts

The perception of environmental damage is not an objective process. It depends both on the availability of knowledge and its evaluation in terms of economic and political criteria. Cultural attitudes to natural resources and man's place in the natural world are important.

Major Anglo-German differences are apparent in this area. German acid-rain policy was formulated in response to both observed and anticipated forest damage. Precautionary environmental policy[1] of this type thrives on uncertainty and risk rather than scientific certainty. Precaution is a philosophical rather than a scientific principle, driving policy-makers to take out environmental insurance policies.

The precautionary approach is deeply embedded in German culture and in the political and legal systems. Remedial policies and legal instruments were adopted between 1982–4 in the absence of established scientific evidence linking forest damage to air pollution. The tentative theory that the long-range transport of sulphur dioxide (SO_2) and nitrogen oxides (NO_x) was to blame was adopted surprisingly quickly by environmentalists, the media, the public and leading politicians. The prompt and energetic policy response resulted from the compatibility of emission abatement strategies with other policy goals.

The fact that forest damage is now known to have been caused by a more subtle mix of factors, with air pollution implicated to an uncertain degree, has not led to any regret about the massive investment in cleaning up power stations. Similar policies with equally large energy, technological and economic implications are now being promoted with reference to the problem of climate change.

Britain's resistance to emission controls has repeatedly been justified on grounds described as scientific.[2] However, whether science was the rationale for established policy or merely the public justification is less clear. The scientific study of air pollution was not well supported in Britain during the 1970s[3] and few institutions were actively looking for environmental damage which would stimulate policy. Even when evidence of foreign damage began to accumulate, little trust was put in the scientific work carried out abroad. Some British science was aimed at cross-checking work carried out elsewhere and many

British scientists worked directly either for the energy industries or for the government. Their work tended to be used to defend the British reluctance to reduce acid emissions.

The initial British response to acid rain was shaped, not by environmental damage at home, but by the allegation of foreign damage caused by trans-boundary air pollution. This is because, in part, British environmental interests tend to be focused on landscape rather than air quality. A vigorous British response to acid rain might have been conceivable if cherished environmental assets, particularly those located in Southern England, had been threatened. The major concern about the loss of broad-leaved trees during the storms of October 1987 and the impacts of constructing a Channel Tunnel rail link to London illustrate this point. In particular, tree damage through air pollution is unlikely to provoke major public concern because mono-culture forestry is an unpopular form of land use in Britain.

The perception of damage from acid deposition in Britain therefore remains much weaker than in Germany. The Timber Foundation claims that not a single British tree has been killed by acid rain and the Forestry Commission, which manages most of Britain's forest plantations, is still sceptical about any direct link between forest damage and air pollution.

Science has turned out to be the servant rather than the master in the making of acid-rain policy in both Britain and Germany, although the nature of its role has been very different. British policy was less active because policy was understood to be driven primarily by the scientific assessment of environmental damage. More explicit account was taken of factors other than scientific evidence in the German policy-making process.

The role of technology

The role played by technology, and technology 'forcing' in the determination of environmental policy objectives and instruments is an important theme in the Anglo-German comparison. Environmental regulation can be used as 'a positive means of technical control, as a means for directing technical change'.[4] It can stimulate investment, alter the demand for specific products and the price structure of competing resources. However, only if the necessary institutional underpinning is in place can account be taken of these potentially positive relationships between regulation, technological innovation and the stimulation of economic and commercial activity.

As long as environmental damage is judged to be tolerable by government there will be little demand for regulation and thus technological or behavioural change. Instead, industrial arguments based on short-term cost calculations will prevail, as was the case in Britain. Change there did not come easily because the traditional style of pollution control is rooted in the pragmatic, incremental policy style of which the UK is proud, but which is now challenged by Europe. During the 1980s, British research into abatement and new power generation technologies was largely in the hands of the nationalized energy industries which had no immediate interest in the abatement of acid emissions.[*]

In Germany, a high value is traditionally placed on technology. R&D is well funded through the Federal Ministry for Research and Technology. The anticipatory approach to environmental policy encouraged the pursuit of wider German technology policy objectives. Institutional arrangements, including the relative weakness of energy interests at the Federal level, further aided this policy. Independent engineering organizations, e.g. the Society of German Engineers (VDI) and the Technical Inspection Offices (TUVs), enjoy high status and can influence policy through the advice which they provide to BMU, the Federal Environment Office (UBA) and the Länder environment ministries.

The German response was also different because there was a perception that the country was engaged, notably with Japan, in an economic battle in which technology is a principal weapon. There is a search for means to encourage technological change and innovation which encompasses the search for cleaner technologies and less toxic materials, and the widespread use of better pollution abatement equipment and advanced monitoring devices.[5]

The degree of emphasis given to the development of cleaner technology is a political choice which depends on government's perception of both its economic goals and its own broader role in society. If there is a technological weakness, or if the commercial perspective is either missing or defensive, then science will probably be asked to fulfil a defensive role.

Economic and commercial impacts

The major investments often needed to reduce emissions establish an obvious link between environmental and economic policy. Both the magnitude of control costs and the source of funding help to determine the acceptability of emissions abatement. While the polluter pays principle, accepted by both Britain and the FR Germany, theoretically determines who will pay for emission controls, the complexity of utility regulation, patterns of ownership and broader government-industry relationships introduce practical complexities.

Britain's acid-rain difficulties stemmed from the potential impact of EC emissions regulation on its nationalized electricity industry, characterized by its dependency on coal and a weak nuclear sector. The cost of retrofitting existing power plants with flue gas desulphurization would have had a major impact on public finances without any apparent compensatory benefits. HM Inspectorate of Pollution and its predecessor bodies responsible for regulating atmospheric emissions were relatively weak and were not prepared to force technical change.[6]

In the 1980s, the British government hoped to transform the economy through unleashing market forces and, in the energy sector, through privatizing the nationally owned monopolies and opening them up to competition. The dominant policy theme was deregulation, within which technology-forcing did not appear an attractive option.

Germany, however, had a stronger regulatory tradition and technology forcing was an acceptable policy aim. The environmental cost burdens put on coal and lignite also promised to make nuclear power more cost-effective and socially acceptable.[7]

The polluting industries in Germany were not, initially, any keener than their British cousins to implement costly measures, especially retrospective requirements on existing power stations. They were compelled to reduce emissions because of the wider perception that benefits for the national economy would follow and competitors would have to bear similar cost burdens through the eventual harmonization of standards at the European level. It has recently been argued in Germany that the early resistance to environmental protection on cost grounds was not a response to a real problem, but a matter of perception.[8]

For the car industry, German policy priorities were to protect export markets, contain Japanese imports and maintain a reputation of well-engineered, high-performance cars. These challenges were perceived to be best met by moving over as rapidly as possible to clean cars equipped with three-way catalytic convertors. Trade and investment implications favoured higher environmental standards.

Viewed from this perspective, pollution abatement initiatives and the considerable associated costs cannot be seen simply as luxuries which 'only the Germans can afford' – a view sometimes heard in Britain. They were part of a rational strategy for 'ecological modernisation'[9] which will, if anything, widen the economic gap between the FR Germany and the UK.

Policy explanations

Features of the decision-making systems

While scientific knowledge, technological capability and economic concerns were, in different measures, important components of the policy debates in Britain and Germany, decisions were very much determined by the cultural and institutional contexts within which the debates took place. Six primary factors which were responsible for the divergence in Anglo-German policies have been identified:

1   attitudes to the environment and environmental protection;
2   the structure and forcefulness of the environmental movements;
3   the nature of political and administrative structures and processes;
4   levels of wealth and styles of economic management which resulted in different perceptions of the benefits and costs of environmental control expenditure;
5   fuel competition, ownership, structure and patterns of regulation in the energy sector; and
6   the vigour of the air pollution control systems.

Clearly, there are also objective differences in the extent to which acid emissions have caused environmental damage in the two countries. However, the six contextual factors which have been identified determined not only the weights attached to the various consequences of policy decisions but also the constellations of interests which supported different policies.

Each one of these themes points towards greater environmental activism in the FR Germany and less in Britain. It is not therefore possible to identify any single factor which leads to precautionary air pollution policy in Germany and reactive policy in the UK. As argued below, it was the conjunction of the six themes which defined the policy space within which top-level environmental policy decisions were made.

Cultural differences

German language and culture tend to invite a more positive approach to air-pollution control than in the case in Britain. The mere presence of pollutants, as opposed to their impacts on ecosystems, is sufficient to stimulate a desire for mitigative action. German culture and history may also create a greater predisposition towards pessimism and anxiety which may, in turn, encourage a more dramatic public response to perceived threats.

Whereas acid rain posed no major threats to cherished environmental assets in Britain, forests, apparently threatened by air pollution, play an important role in German culture as well as being of considerable economic value. This partly explains the hasty legislative response to the forest dieback problem in Germany in the early 1980s, but not the longer-term emphasis on pro-active environmental policies.

At that time, environmental concern was a potentially unifying theme for a nation divided over other major policy questions, such as the stationing of US missiles on German soil and the development of nuclear power. In addition, self-doubts about the perceived sterility of the economic miracle helped to create a fertile ground for arousing public concern which, in turn, encouraged the growth of the green movement.

The environmental movements

There are substantial differences between the environmental movements in Britain and Germany. Deep green thinking plays a smaller role in the British movement which is determinedly non-party political and largely operates 'within the system'. In Germany, a small but influential political party has grown out of a radical green movement born in opposition to nuclear power and sceptical about the value of economic growth for its own sake.

Given the German electoral system, the Green Party has had a major impact on national politics and has forced the established parties to compete with each other over environmental policy. Enhanced attention to environmental protection has been the official response to what started out as a much more fundamental challenge to the status quo.

In Britain, there was little party-political competition over environmental issues during the 1980s, though this is now beginning to emerge. On acid rain, the environmental movement's attempts to put direct pressure on the government to institute remedial action were largely ineffective.

Party politics and government

The generally more intensive degree of party political competition in the FR Germany has created a greater degree of interaction between public opinion and the political system on environmental questions. Because, in the Federal system, elections are virtually continuous, the political parties are much more sensitive to changing public moods. There are, vice versa, greater incentives for government to influence public opinion, for example through the official 'Save Our Forest Federal Action Programme'.

Under the two-party system in Britain, the dominant position of the party in power, which was even more marked than normal during the 1980s, allows the goverment to insulate itself from public opinion, making it more sensitive to narrower expressions of economic or ideological interest. This factor is reinforced by the greater geographical and institutional centralization of power.

The federal structure of government in Germany leads to a broad, if cumbersome, consensus-building decision-making process which favoured environmental issues. Competition between the Federal and Land governments in Germany for jurisdiction over environmental issues also created more political energy, driving forward environmental activism.

Economic management

Both the relative strengths of the British and German economies, and the styles of economic management, affected the outcomes of the respective national debates on acid emission controls. In the FR Germany, the limited number of macro-economic instruments available to the Federal government led to advantages being perceived in the stimulation of environmental protection investment through regulation. There was also a strong perception that the international market for environmental protection equipment would grow and would represent a major opportunity for German industry. Domestic markets would provide a strong base for market growth.

In Britain, the costs of a major investment programme to clean up power-station emissions would have been incurred by the state-owned electricity supply industry at a time of tight restrictions on public expenditure. For Britain, the abatement of acid emissions represented a threat rather than an opportunity.

Energy sector structure

Fundamental differences in ownership and structure of the electricity supply industries have been of great importance in influencing air-pollution policies in Britain and Germany, although some similarities exist, both countries being heavily dependent on high-cost domestically produced coal.

The UK power industry, in national ownership throughout the 1980s, has been tightly bound to central government and was, until very recently, permitted to play a major role in the determination of acid-rain policy. Until 1986, it resolved its potentially conflicting duties to maintain an economic and efficient electricity supply system and to take note of the environmental impacts of its activities by opposing acid-rain controls. Privatization of the power industry is changing this context, though it is not clear whether pro-active environmental policy will become more or less easy to pursue.

The more numerous German utilities function as private sector companies but have a complex set of relationships with government at the Federal, Land and municipal levels through both regulation and share ownership. Perceived regional injustices in the system of utility-administered financial aid for the hard coal industry were a factor encouraging environmental controls which promised to weaken coal's market position and assist other energy sources, notably nuclear power. For various political reasons, the nuclear power argument in favour of acid-rain controls was not advanced in the UK.

Air pollution control

The German air-pollution control regime is considerably more legalistic than that in the UK, requiring the clear articulation of broad philosophies and goals. The more pro-active air pollution system has been created partly by Federal–Land tensions. In the late 1970s, the utilities sought uniform Federal legislation to avoid what they saw as the arbitrary application of air-pollution guidelines by the Länder authorities.

In Britain, much of the responsibility for interpreting environmental law falls to the discretion of regulators who were relatively weak and unassertive during the late 1970s and 1980s. The Industrial Air Pollution Inspectorate did not press for acid-rain controls which were initiated, in the end, by the Central Electricity Generating Board.

The convergence of policy influences

The strong conclusion reached from this analysis of environmental decision-making is that no single factor can be isolated which explains why countries choose any specific course of action. The Anglo-German comparison shows that it is necessary to call upon a variety of underlying influences – attitudes, ideology, party political competition, institutional arrangements, corporate interests and macro-economic factors – in addition to the objective assessment of environmental damage and remedial costs.

International pressures may also be an important influence on environmental policy. However, international pressure, unless accompanied by similar domestic demands, is likely to stimulate an initially defensive response from countries accused of transboundary pollution. This was true in Germany, as much as in the UK, until the discovery of forest dieback added the necessary political energy to the debate.

In the UK, the considerable support in the wider political system (outside the government) for acid-rain controls in 1984–5 was not reflected in the population at large. Other political factors tended to militate against action, until international pressures, coupled with a rapidly evolving domestic agenda, converged to permit the formation of a more limited acid emissions abatement plan.

However, the implementation plan for emissions abatement developed during 1990 shows the continuing linkages between the environment and other policy concerns. The successful international pressure on the UK to moderate its policy came not from Scandinavia, where there is conclusive evidence of environmental damage to which the UK contributes, but from the FR Germany which is relatively unaffected by UK emissions and from the EC where one of the major concerns is uniformity of control costs.

The relative importance of the different factors explaining acid-rain policy is virtually impossible to judge. In Germany, wide public concern and its impact on the political system appear to have been the factors which triggered policy action from a political and regulatory system which was predisposed to act.

However, the German case study also shows that the different factors underlying policy are interdependent. From one perspective, public concern about forest damage in Germany led political action. On the other hand, sensing the opportunity to further a variety of policy objectives, the Federal government clearly participated, with academics and the media, in the stimulation of public opinion. These two processes were simultaneous. This phenomenon of government-led environmentalism emerged in Britain in the late 1980s in the context of the climate change debate. However, the very existence of a latent public concern about air pollution in Germany in the early 1980s is significant in itself.

In Britain, even this latent public concern about acid rain appears to have been absent. The fact that the government finally decided to accept an emissions abatement policy is a testimony to the fact that public interest is not a necessary condition for policy action and to the existence of limits to British sovereignty in policy fields where the EC has taken an active role.

The analysis also casts light on the process of European 'policy integration', the process by which the EC anticipates environmental objectives becoming a component of other policies. In the past, it appears that environmental measures have had to be compatible with wider policy concerns, such as energy and economic management. Reversing this compatibility requirement to ensure that other policies are compatible with environment goals will be a major challenge which can be met only at a political level.

Policy evaluation

While explaining environmental policy differences is, in principle, a straight-forward empirical task, the evaluation of the success or otherwise of policy is essentially subjective. However, it is possible to identify a number of criteria against which Britain's and Germany's policies on acid rain might be judged. The weight attached to each of these criteria is very much a matter of judgement. No judgement is made here about whether British or German policy was 'right' or 'wrong'. Though acid rain may have been one of the 'conspicuous failures' of British environmental policy during the 1980s,[10] it is concluded here that the UK government acted rationally in the context of its wider political priorities.

The criteria discussed in this section are:

1   environmental impacts;
2   cost implications;
3   impacts on directly affected industrial sectors;
4   wider economic impacts, including effects on technological capability;
5   domestic political impacts; and
6   effects on international relations.

Environmental impact

As far as acid emissions from stationary plant are concerned, German policy has clearly achieved much more than the UK, since power plant SO_2 emissions have already declined by 80 per cent and NO_x emissions are expected to fall by a similar amount by the mid-1990s. UK emissions are unlikely to decline to this extent until well beyond the end of the century. For motor vehicles, technical improvements in vehicle emissions are likely to be offset by increases in traffic levels in both countries.

In terms of air quality, the picture is less clear. Urban air quality is determined by a mixture of short-range and distant emission sources. The more rapid clean-up of motor vehicles in the FR Germany coupled with the dramatic reduction in power plant emissions is likely to lead to enhanced air quality. However, having a continental climate, Germany is affected by transboundary air pollution to a greater extent than is the UK. A clean-up in neighbouring countries to the east, and in the new Eastern Länder in particular, would be needed to enhance air quality significantly throughout the Federal Republic.[11] The UK is largely responsible for its own air quality and modest improvements may be expected in coming years. In both countries, background air quality, away from population centres, is determined primarily by emissions from distant sources.

The impacts of air pollution on sensitive ecological targets will be affected by the policies of the 1980s. The effect of Germany's clean-up programme on forest health is not clear, but subsequent scientific analysis shows that it is certain to be far less than anticipated in 1982–3. Forest health had stabilized, and was even improving slightly, before power-station emissions began to decline substantially. The longer-term effect of reduced motor vehicle emissions, if achieved, may make a more positive contribution.

Ironically, had Britain instituted a clean-up programme on the German scale it may have been possible to arrest certain types of environmental damage. Based on the conclusions of the UK Acid Waters Review Group, an 80 per cent reduction in power station SO_2 emissions would go much of the way towards preventing the further deterioration of acid waters in remote parts of the UK. There would be a smaller effect in Scandinavia where acid deposition damage, caused partly by the UK, is well understood. German emission reductions between 1980 and 1988, as shown in Table 1.1 and 1.2, have reduced SO_2

deposition in Norway and Sweden by 5–6 per cent. The effect in the more sensitive Southern parts of Scandinavia is likely to have been higher.

Cost implications

The cost implications of British and German policies are well established. The German power station clean-up programme is likely to cost DM 21bn between 1983 and 1993, with most of the cost incurred by 1988. Almost all of this has been borne by electricity consumers, except for certain projects of an innovative nature, part of the costs of which have been borne by the Federal government. The cost of the clean-up is equivalent to 1.3 Pf/kWh,[12] broadly comparable to the size of the Kohlepfennig ('coal penny') subsidy. However, the impact on electricity prices is less easy to identify because of the nature of regulatory controls and the complex cross-subsidies relating to coal use.

On present plans, active SO_2 and NO_x retrofit measures in the UK (FGD and low-NO_x burners) will cost around £1.6bn, less than a quarter of the sum expended in the FR Germany. A significant proportion of the abatement required by the LCP Directive will be met by burning low sulphur-imported coal or accelerating the installation of combined cycle gas turbine power stations. To the extent that these developments would have taken place without environmental controls, they may be regarded as costless.

The largest part of the UK's abatement costs will be borne by the taxpayer in the form of diminished proceeds from the government's flotation of the electricity-generating companies in early 1991. To this extent, the polluter-pays principle will not apply because, given the new spot market for electricity, there is no mechanism for recovering costs from consumers.

Energy sector impacts

There is now some concern in Germany that the installation of FGD on 37 000 MW_e of power plant capacity has created an inflexible and expensive supply system. Most of the capacity has been installed at hard coal-fired plant which might have a limited life if EC and internal pressures to reduce support for the heavily subsidized coal industry succeed.

On the other hand, the very existence of the FGD installations makes the burning of hard coal and lignite environmentally acceptable and may serve to prolong the life of some of the existing power stations. Given that new power generation technologies, such as combined cycle gas turbine plant, are rapidly being adopted in other parts of Europe, there is however a risk that large parts of the German power sector may become technologically obsolescent. It is an irony that the pursuit of high environmental standards and state-of-the-art abatement technology might yet help to freeze basic power generation technology.

The fact that 12 000 MW_e of coal-fired plant was closed as a result of the domestic LCP Regulation has been of little importance because of the existence of excess generation capacity. In spite of the way in which emission controls have improved the cost-effectiveness of nuclear power relative to fossil fuels, the power industry's desire to build more nuclear power stations has not been fulfilled and public acceptability of the technology has not been improved. The industry now argues that nuclear power is needed to help reduce greenhouse gas emissions.

In Britain, where privatization is removing government control over fuel procurement for electrcity supply, the less ambitious abatement programme is having quite different effects. The regulatory device used to implement EC requirements, the company SO_2 emissions total, which allows each generator to choose a least-cost abatement strategy, provides a incentive which is entirely different from that of a plant-specific emission limit.

The result is that compliance in the UK is likely to come from a mixture of FGD, low sulphur imported coal and the accelerated introduction of combined cycle gas turbine plant. The environmental controls are thus contributing to reductions in the output of British coal, reinforcing the generators' desire to diversify sources of fuel supply and encouraging the more rapid adoption of new generation technology. It has also been suggested that the compliance regime in the UK has helped to reinforce the domination of the CEGB's successor companies in the electricity market.[13]

Wider economic effects

The broader German policy thrust of cleaning up power stations, other stationary sources and motor vehicles has undoubtedly created a greater technological and industrial capability in the growing market for environmental control equipment. Given the EC's 1992 objectives, it is also the case that suppliers of power station equipment are operating in a market which is increasingly European rather than national in character. German manufacturers, such as Siemens, have been successful in tendering for the construction of combined cycle gas turbine plant in the UK in the absence of large domestic markets. Equally, however, some British companies selling products which contribute to environmental protection have succeeded without having a home market. For example, Davy McKee has secured orders for the construction of FGD plant in the FR Germany while Johnson Matthey has been uniquely placed to supply catalysts for catalytic convertors.

The German and EC objective of 'spreading the misery' of abatement costs in order to equalize terms of international competition may be regarded as a failure, at least in the short term, since environmental control costs have influenced electricity prices in Germany to a greater extent than anywhere else. In the longer term, as power plants are replaced and harmonized environmental standards are applied, the effects may become more even.

This fact that British environmental controls were paid for by the taxpayer has wider macro-economic implications given that political concern over the level of public expenditure has re-emerged in 1990.

Domestic political impacts

For government and the established political parties in Germany, environmental controls have served a valuable function. They may have prevented the Green Party from greater electoral successes and, conceivably, from holding the balance of political power. A red–green coalition was a real fear of the Union parties in the early 1980s. In a broader sense, the environmental programme also served a healing function in the Federal Republic as it distracted attention away from fundamental political divisions.

In view of the relatively low level of public interest, the domestic impact of British acid-rain policy has been minimal. Lead in petrol was the only air-pollution topic to be the subject of vigorous debate during the 1980s. Public debate was provoked in 1990, however, by the discussion of plans for the implementation of air pollution controls in power generation and, in particular, the potential effects on the British coal industry.

International relations

British acid-rain policy has, during most of the 1980s, had a negative effect on links with the Scandinavian countries with whom the UK, in general, enjoys the most cordial relations. The strain on relationships has diminished since the agreement of the LCP Directive, since Scandinavian concern is now focused more on Eastern Europe. Britain's obdurate resistance to the LCP Directive and US-type vehicle emission standards has also provoked some resentment in the EC, though this has to be seen within the wider context of difficult UK–EC relations during the 1980s and much more fundamental disagreements over Community budget contributions.

The FR Germany's policies have earned itself a place as one of three 'green' Member States of the EC (the others being the Netherlands and Denmark) which are in the driving seat as far as Community environmental policy is concerned. On the other hand, the force and inflexibility of German support for the LCP Directive has been a source of some resentment in parts of the British political community.

Conclusions

The Anglo-German comparison shows that the environmental policies preferred by different countries are unlikely to coincide unless there are strong similarities in terms of living standards, cultural attitudes, economic characteristics and institutional structures. Such a convergence is not a realistic possibility given the diversity of the nations of Europe. The need for drawn-out negotiations and continuous adjustments within the EC, as well as at a wider European level, remains. Environmental policy, intrinsically linked to the issue of commercial competitiveness, forms part of a wider power struggle.

Britain is now engaged in a process of adaptation to European and domestic environmental pressures, while still facing difficulties with the implementation of EC legislation which it has already accepted. The precautionary, technology-led formulation of environmental policy in Brussels, stimulated by the FR Germany above all, poses considerable challenges for the UK which has, in the past, taken a reactive approach underpinned by rigorous scientific scepticism.

Nevertheless, the rhetoric of recent high-level British statements on climate change is considerably more precautionary than that adopted with regard to any previous environmental problem. However, the key policy question is whether Britain, after years of neglect, has the institutional capability and the political will to put ambitious, precautionary policies into effect. If it does not, intergovernmental wrangling over energy and environmental policy, of the kind which characterized the acid-rain debate in the 1980s, will also be a feature of the 1990s.

The situation in Germany, on the other hand, is now such that the environmental imperative as perceived by large and powerful sections of society is even more pressing than that recognized in EC policy.[14] There is a risk that the German public, led by the green movement, may turn away from a Europe the primary aim of which is commercial and technological competition with Japan. The future debate, which must now encompass the development of energy resources and environmental controls in Eastern Europe, promises to be exciting, challenging and of enduring interest.

Notes

1. K. von Moltke (1987), *The Vorsorgeprinzip in West German Environment Policy*, paper submitted to the UK Royal Commission on Environmental Pollution for its 12th Report, mimeo.
2. T. O'Riordan (1989), Air Pollution Legislation in the European Community, *Atmospheric Environment*, Vol. 23, No. 2.
3. Environmental science attracted only 17 per cent of all research funding for basic research in 1986 (*New Scientist*, 19 August 1989). Over half of public R&D funds still go to defence, see R. Williams (1988), UK Science and Technology: Policy, Controversy and Advice, *The Political Quarterly*, Vol. 65, No. 2; and various contributors to The Royal Society (May 1989), *Science and Public Affairs*, Vol. 4, No. 1.
4. A. Irwin and P. Bergragt (1989), Rethinking the relationship between environmental regulation and industrial innovation: the social negotiation of technical change, in *Technology Analysis and Strategic Management*, Vol 1., No. 1; also P. Patel and K. Pavitt (March 1988), *Technological Activities in FR Germany: Differences and Determinants*, DRC Discussion Paper No 58, SPRU, Brighton.
5. R. van Tulden and G. Junne (1989), *European Multinationals in Core Technologies*, Wiley.
6. T. O'Riordan and A. Weale, Administrative reorganisation and policy change: the case of HMIP, *Journal of Public Administration*, 67,3, Autumn 1989.
7. P. Wiedemann and H. Jungermann (March 1989), Energy and the public, *Arbeiten zur Risiko-Kommunikation Heft 6*, Jülich.
8. E. Müller (17 November 1989), Sozial-liberale Umweltpolitik: Von der Karriere eines neuen Politibereichs, *Aus Politik und Zeitgeschichte*, Beilage zu Parliament, Bonn.
9. H. Maarten and M. Schwarz (September 1989), Acid rain and the cultural climate:

what is the problem, *Congress on Scientific Controversies and Political Decisions concerning the Environment*, Arc-et-Senans, France.

10. M. Holdgate in Centre for Economic and Environmental Development (July/August 1989), *CEED Bulletin No 24*, London.

11. The combined SO_2 emissions of East and West Germany are higher in per capita terms than those in the UK.

12. B. Schärer (1989), Economic impact of flue gas cleaning at power plants in the Federal Republic of Germany, in L.J. Brasser and W.C. Mulder (eds), *Man and His Ecosystem: Proceedings of the 8th World Clean Air Congress 1989*, Elsevier, Amsterdam, Vol. 5, pp. 85–90.

13. J. Skea (1990), Memorandum in House of Commons Energy Committee, *The Flue Gas Desulphurisation Programme*, report, Session 1989–90, HC, HMSO, London.

14. Political pressures may demand the implementation of emerging slogans, such as 'ecological perestroika' (SPD), *Zukunftsaufgabe ökologischer Umbau*, Bonn, 1989), the 'ecological market economy', and 'man's responsibility for the Creation' (CDU, Dokumentation 29, *Unsere Verantwortung für die Schöpfung*, Bonn 1989). For further insights on German attitudes, see von Moltke's review of Mayer-Tisch, Die verseuchte Landkarte, in *Environmental Affairs*, (Winter 1989).

Index

290